STUDENT'S SOLUTIONS MANUAL

JAMES LAPP
Colorado Mesa University

ELEMENTARY STATISTICS
TWELFTH EDITION

Mario F. Triola

Dutchess Community College

Boston Columbus Indianapolis New York San Francisco Upper Saddle River
Amsterdam Cape Town Dubai London Madrid Milan Munich Paris Montreal Toronto
Delhi Mexico City São Paulo Sydney Hong Kong Seoul Singapore Taipei Tokyo

ISBN-13: 978-0-321-83792-9
ISBN-10: 0-321-83792-4

2 3 4 5 6 V011 17 16 15 14

www.pearsonhighered.com

CONTENTS

Chapter 1: Introduction to Statistics

Section 1-2

1. Statistical significance is indicated when methods of statistics are used to reach a conclusion that some treatment or finding is effective, but common sense might suggest that the treatment or finding does not make enough of a difference to justify its use or to be practical. Yes, it is possible for a study to have statistical significance but not a practical significance.

3. A voluntary response sample is a sample in which the subjects themselves decide whether to be included in the study. A voluntary response sample is generally not suitable for a statistical study because the sample may have a bias resulting from participation by those with a special interest in the topic being studied.

5. There does appear to be a potential to create a bias.

7. There does not appear to be a potential to create a bias.

9. The sample is a voluntary response sample and is therefore flawed.

11. The sampling method appears to be sound.

13. Because there is a 30% chance of getting such results with a diet that has no effect, it does not appear to have statistical significance, but the average loss of 45 pounds does appear to have practical significance.

15. Because there is a 23% chance of getting such results with a program that has no effect, the program does not appear to have statistical significance. Because the success rate of 23% is not much better than the 20% rate that is typically expected with random guessing, the program does not appear to have practical significance.

17. The male and female pulse rates in the same column are not matched in any meaningful way. It does not make sense to use the difference between any of the pulse rates that are in the same column.

19. The data can be used to address the issue of whether males and females have pulse rates with the same average (mean) value.

21. Yes, each IQ score is matched with the brain volume in the same column, because they are measurements obtained from the same person. It does not make sense to use the difference between each IQ score and the brain volume in the same column, because IQ scores and brain volumes use different units of measurement. For example, it would make no sense to find the difference between an IQ score of 87 and a brain volume of 1035 cm^3.

23. Given that the researchers do not appear to benefit from the results, they are professionals at prestigious institutions, and funding is from a U.S. government agency, the source of the data appears to be unbiased.

25. It is questionable that the sponsor is the Idaho Potato Commission and the favorite vegetable is potatoes.

27. The correlation, or association, between two variables does not mean that one of the variables is the cause of the other. Correlation does not imply causation.

29. a.   The number of people is $(0.39)(1018) = 397.02$

    b.   No. Because the result is a count of people among 1018 who were surveyed, the result must be a whole number.

    c.   The actual number is 397 people

    d.   The percentage is $\dfrac{255}{1018} = 0.25049 = 25.049\%$

31. a.   The number of adults is $(0.14)(2302) = 322.28$

    b.   No. Because the result is a count of adults among 2302 who were surveyed, the result must be a whole number.

    c.   The actual number is 322 adults.

    d.   The percentage is $\dfrac{46}{2302} = 0.01998 = 1.998\%$

33. Because a reduction of 100% would eliminate all of the size, it is not possible to reduce the size by 100% or more.

35. If foreign investment fell by 100% it would be totally eliminated, so it is not possible for it to fall by more than 100%.

37. Without our knowing anything about the number of ATVs in use, or the number of ATV drivers, or the amount of ATV usage, the number of 740 fatal accidents has no context. Some information should be given so that the reader can understand the rate of ATV fatalities.

39. The wording of the question is biased and tends to encourage negative response. The sample size of 20 is too small. Survey respondents are self-selected instead of being selected by the newspaper. If 20 readers respond, the percentages should be multiples of 5, so 87% and 13% are not possible results.

Section 1-3

1.   A parameter is a numerical measurement describing some characteristic of a population, whereas a statistic is a numerical measurement describing some characteristic of a sample.

3.   Parts (a) and (c) describe discrete data.

5.   Statistic   17. Discrete

7.   Parameter   19. Continuous

9.   Parameter   21. Nominal

11. Statistic   23. Interval

13. Continuous   25. Ratio

15. Discrete   27. Ordinal

29. The numbers are not counts or measures of anything, so they are at the nominal level of measurement, and it makes no sense to compute the average (mean) of them.

31. The numbers are used as substitutes for the categories of low, medium, and high, so the numbers are at the ordinal level of measurement. It does not make sense to compute the average (mean) of such numbers.

33. a.   Continuous, because the number of possible values is infinite and not countable.

    b.   Discrete, because the number of possible values is finite.

    c.   Discrete, because the number of possible values is finite.

    d.   Discrete, because the number of possible values is infinite and countable.

35. With no natural starting point, temperatures are at the interval level of measurement, so ratios such as "twice" are meaningless.

Section 1-4

1.   No. Not every sample of the same size has the same chance of being selected. For example, the sample with the first two names has no chance of being selected. A simple random sample of (n) items is selected in such a way that every sample of same size has the same chance of being selected.

3.   The population consists of the adult friends on the list. The simple random sample is selected from the population of adult friends on the list , so the results are not likely to be representative of the much larger general population of adults in the United States.

5.   Because the subjects are subjected to anger and confrontation, they are given a form or treatment, so this is an experiment, not an observational study.

7.   This is an observational study because the therapists were not given any treatment. Their responses were observed.

9.   Cluster

11. Random

13. Convenience

15. Systematic

17. Random

19. Convenience

21. The sample is not a simple random sample. Because every 1000th pill is selected, some samples have no chance of being selected. For example, a sample consisting of two consecutive pills has no chance of being selected, and this violates the requirement of a simple random sample.

23. The sample is a simple random sample. Every sample of size 500 has the same chance of being selected.

25. The sample is not a simple random sample. Not every sample has the same chance of being selected. For example, a sample that includes people who do not appear to be approachable has no chance of being selected.

27. Prospective study

29. Cross-sectional study

31. Matched pairs design

33. Completely randomized design

35. Blinding is a method whereby a subject (or a person who evaluates results) in an experiment does not know whether the subject is treated with the DNA vaccine or the adenoviral vector vaccine. It is important to use blinding so that results are not somehow distorted by knowledge of the particular treatment used.

Chapter Quick Quiz

1.   No. The numbers do not measure or count anything.

2.   Nominal

3.   Continuous

4.   Quantitative data

5.   Ratio

6.   False

7.   No

8.   Statistic

9.   Observational study

10   False

Review Exercises

1.   a.   Discrete

     b.   Ratio

     c.   Stratified

     d.   Cluster

     e.   The mailed responses would be a voluntary response sample, so those with strong opinions are more likely to respond. It is very possible that the results do not reflect the true opinions of the population of all costumers.

2.   The survey was sponsored by the American Laser Centers, and 24% said that the favorite body part is the face, which happens to be a body part often chosen for some type of laser treatment. The source is therefore questionable.

3.   The sample is a voluntary response sample, so the results are questionable.

4.   a.   It uses a voluntary response sample, and those with special interests are more likely to respond, so it is very possible that the sample is not representative of the population.

     b.   Because the statement refers to 72% of all Americans, it is a parameter (but it is probably based on a 72% rate from the sample, and the sample percentage is a statistic).

     c.   Observational study.

5. a. If they have no fat at all, they have 100% less than any other amount with fat, so the 125% figure cannot be correct.

    b. The exact number is $(0.58)(1182) = 685.56$. The actual number is 686.

    c. $\dfrac{331}{1182} = 0.28003 = 28.003\%$

6. The Gallop poll used randomly selected respondents, but the AOL poll used a voluntary response sample. Respondents in the AOL poll are more likely to participate if they have strong feelings about the candidates, and this group is not necessarily representative of the population. The results from the Gallop poll were more likely to reflect the true opinions of American voters.

7. Because there is only a 4% chance of getting the results by chance, the method appears to have statistical significance. The results of 112 girls in 200 births is above the approximately 50% rate expected by chance, but it does not appear to be high enough to have practical significance. Not many couples would bother with a procedure that raises the likelihood of a girl from 50% to 56%.

8. a. Random

    b. Stratified

    c. Nominal

    d. Statistic, because it is based on a sample.

    e. The mailed responses would be a voluntary response sample. Those with strong opinions about the topic would be more likely to respond, so it is very possible that the results would not reflect the true opinions of the population of all adults.

9. a. Systematic

    b. Random

    c. Cluster

    d. Stratified

    e. Convenience

    f. No, although this is a subjective judgment.

10. a. $0.52(1500) = 780$ adults

    b. $\dfrac{345}{1500} = 0.23 = 23\%$

    c. Men: $\dfrac{727}{1500} = 0.485 = 48.5\%$;

        Women: $\dfrac{773}{1500} = 0.515 = 51.5\%$

Cumulative Review Exercises

1. The mean is 11. Because the flight numbers are not measures or counts of anything, the result does not have meaning.

2. The mean is 101, and it is reasonably close to the population mean of 100.

3. $\dfrac{(247-176)}{6} = 11.83$ is an unusually high value.

4. $\dfrac{(175-172)}{\left(\dfrac{29}{\sqrt{20}}\right)} = 0.46$

5. $\dfrac{\left(1.96^2 \times 0.25\right)}{0.03^2} = 1067$

6. $\dfrac{(88-88.57)^2}{88.57} = 0.0037$

7. $\dfrac{\left((96-100)^2 + (106-100)^2 + (98-100)^2\right)}{(3-1)} = 28.0$

8. $\sqrt{\dfrac{\left((96-100)^2+(106-100)^2+(98-100)^2\right)}{(3-1)}} = \sqrt{28} = 5.3$

9. $0.6^{14} = 0.00078364164$

10. $8^{12} = 68719476736$

11. $7^{14} = 678223072849$

12. $0.3^{10} = 0.0000059049$

Chapter 2: Summarizing and Graphing Data

Section 2-2

1. No. For each class, the frequency tells us how many values fall within the given range of values, but there is no way to determine the exact IQ scores represented in the class.

3. No. The sum of the percentages is 199% not 100%, so each respondent could answer "yes" to more than one category. The table does not show the distribution of a data set among all of several different categories. Instead, it shows responses to five separate questions.

5. Class width: 10.
    Class midpoints: 24.5, 34.5, 44.5, 54.5, 64.5, 74.5, 84.5.
    Class boundaries: 19.5, 29.5, 39.5, 49.5, 59.5, 69.5, 79.5, 89.5.

7. Class width: 10.
    Class midpoints: 54.5, 64.5, 74.5, 84.5, 94.5, 104.5, 114.5, 124.5.
    Class boundaries: 49.5, 59.5, 69.5, 79.5, 89.5, 99.5, 109.5, 119.5, 129.5.

9. Class width: 2.
    Class midpoints: 3.95, 5.95, 7.95, 9.95, 11.95.
    Class boundaries: 2.95, 4.95, 6.95, 8.95, 10.95, 12.95.

11. No. The frequencies do not satisfy the requirement of being roughly symmetric about the maximum frequency of 34.

13. 18, 7, 4

15. On average, the actresses appear to be younger than the actors.

Age When Oscar Was Won	Relative Frequency (Actresses)	Relative Frequency (Actors)
20 – 29	32.9%	1.2%
30 – 39	41.5%	31.7%
40 – 49	15.9%	42.7%
50 – 59	2.4%	15.9%
60 – 69	4.9%	7.3%
70 – 79	1.2%	1.2%
80 – 89	1.2%	0.0%

17. The cumulative frequency table is

Age (years) of Best Actress When Oscar Was Won	Cumulative Frequency
Less than 30	27
Less than 40	61
Less than 50	74
Less than 60	76
Less than 70	80
Less than 80	81
Less than 90	82

19. Because there are disproportionately more 0s and 5s, it appears that the heights were reported instead of measured. Consequently, it is likely that the results are not very accurate.

x	Frequency
0	9
1	2
2	1
3	3
4	1
5	15
6	2
7	0
8	3
9	1

21. Yes, the distribution appears to be a normal distribution.

Pulse Rate (Male)	Frequency
40 – 49	1
50 – 59	7
60 – 69	17
70 – 79	9
80 – 89	5
90 – 99	1

23. No, the distribution does not appear to be a normal distribution.

Magnitude	Frequency
0.00 – 0.49	5
0.50 – 0.99	15
1.00 – 1.49	19
1.50 – 1.99	7
2.00 – 2.49	2
2.50 – 2.99	2

25. Yes, the distribution appears to be roughly a normal distribution.

Red Blood Cell Count	Frequency
4.00 – 4.39	2
4.40 – 4.79	7
4.80 – 5.19	15
5.20 – 5.59	13
5.60 – 5.99	3

27. Yes. Among the 48 flights, 36 arrived on time or early, and 45 of the flights arrived no more than 30 minutes late.

Arrival Delay (min)	Frequency
(−60) − (−31)	11
(−30) − (−1)	25
0 − 29	9
30 − 59	1
60 − 89	0
90 − 119	2

29.

Category	Relative Frequency
Male Survivors	16.2%
Males Who Died	62.8%
Female Survivors	15.5%
Females Who Died	5.5%

31. Pilot error is the most serious threat to aviation safety. Better training and stricter pilot requirements can improve aviation safety.

Cause	Relative Frequency
Pilot Error	50.5%
Other Human Error	6.1%
Weather	12.1%
Mechanical	22.2%
Sabotage	9.1%

33. An outlier can dramatically affect the frequency table.

Weight (lb)	With Outlier	Without Outlier
200 − 219	6	6
229 − 239	5	5
240 − 259	12	12
260 − 279	36	36
280 − 299	87	87
300 − 319	28	28
320 − 339	0	
340 − 359	0	
360 − 379	0	
380 − 399	0	
400 − 419	0	
420 − 439	0	
440 − 459	0	
460 − 479	0	
480 − 499	0	
500 − 519	1	

Section 2-3

1.   It is easier to see the distribution of the data by examining the graph of the histogram than by the numbers in the frequency distribution.

3.   With a data set that is so small, the true nature of the distribution cannot be seen with a histogram. The data set has an outlier of 1 minute. That duration time corresponds to the last flight, which ended in an explosion that killed seven crew members.

5.   Identifying the exact value is not easy, but answers not too far from 200 are good answers.

7.   The tallest person is about 108 inches, or about 9 feet tall. That tallest height is depicted in the bar that is farthest to the right in the histogram. That height is an outlier because it is very far from all of the other heights. The height of 9 feet must be an error, because the height of the tallest human ever recorded was 8 feet 11 inches.

9.   The digits 0 and 5 seem to occur much more than the other digits, so it appears that the heights were reported and not actually measured. This suggests that the results might not be very accurate.

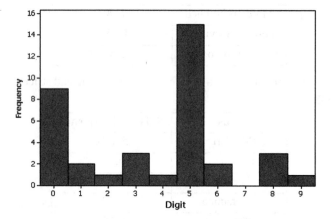

11. The histogram does appear to depict a normal distribution. The frequencies increase to a maximum and then tend to decrease, and the histogram is symmetric with the left half being roughly a mirror image of the right half.

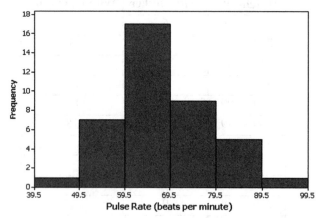

13. The histogram appears to roughly approximate a normal distribution. The frequencies increase to a maximum and then tend to decrease, and the histogram is symmetric with the left half being roughly a mirror image of the right half.

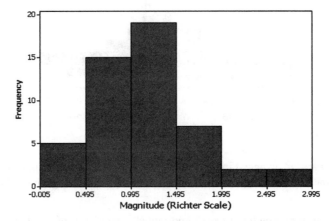

15. The histogram appears to roughly approximate a normal distribution. The frequencies increase to a maximum and then tend to decrease, and the histogram is symmetric with the left half being roughly a mirror image of the right half.

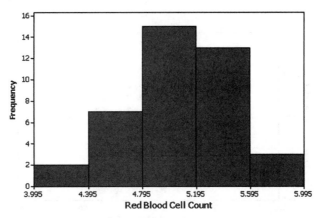

17. The two leftmost bars depict flights that arrived early, and the other bars to the right depict flights that arrived late.

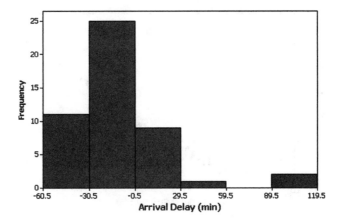

19. The ages of actresses are lower than those of actors.

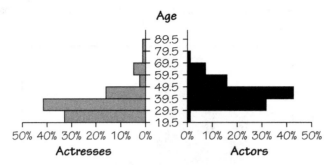

Section 2-4

1. In a Pareto chart, the bars are arranged in descending order according to frequencies. The Pareto chart helps us understand data by drawing attention to the more important categories, which have the highest frequencies.

3. The data set is too small for a graph to reveal important characteristics of the data. With such a small data set, it would be better to simply list the data or place them in a table.

5. Because the points are scattered throughout with no obvious pattern, there does not appear to be a correlation.

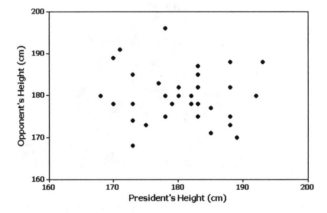

7. Yes. There is a very distinct pattern showing that bears with larger chest sizes tend to weigh more.

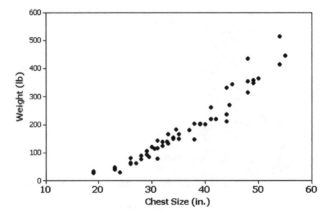

9. The first amount is highest for the opening day, when many Harry Potter fans are most eager to see the movie; the third and fourth values are from the first Friday and the first Saturday, which are the popular weekend days when movie attendance tends to spike.

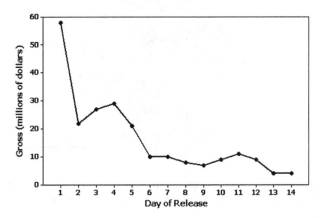

11. Yes, because the configuration of the points is roughly a bell shape, the volumes appear to be from a normally distributed population. The volume of 11.8 oz. appears to be an outlier.

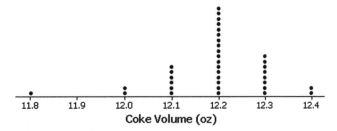

13. No. The distribution is not dramatically far from being a normal distribution with a bell shape, so there is not strong evidence against a normal distribution.

    4 | 5
    5 | 3 3 3 5 5 7 9
    6 | 1 1 1 6 7
    7 | 1 1 1 1 5 5 6 8
    8 | 4

15.

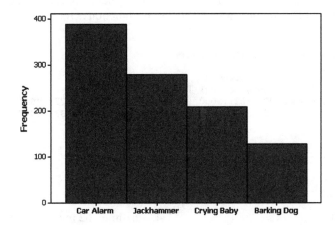

17.

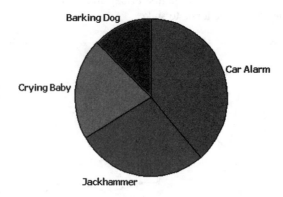

19. The frequency polygon appears to roughly approximate a normal distribution. The frequencies increase to a maximum and then tend to decease, and the graph is symmetric with the left half being roughly a mirror image of the right half.

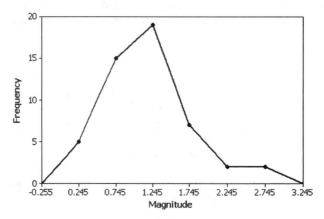

21. The vertical scale does not start at 0, so the difference is exaggerated. The graphs make it appear that Obama got about twice as many votes as McCain, but Obama actually got about 69 million votes compared to 60 million to McCain.

23. China's oil consumption is 2.7 times (or roughly 3 times) that of the United States, but by using a larger barrel that is three times as wide and three times as tall (and also three times as deep) as the smaller barrel, the illustration has made it appear that the larger barrel has a volume that is 27 times that of the smaller barrel. The actual ratio of US consumption to China's consumption is roughly 3 to 1, but the illustration makes it appear to be 27 to 1.

25. The ages of actresses are lower than those of actors.

```
                           Actresses   Actors
          9999999888877776666 55554421 | 2 | 9
  998888776555 5555444333333222111000 | 3 | 0012224445 5666777788888999
              9655322111110 | 4 | 00011111122222233334445 55677788999
                         40 | 5 | 0112222346677
                       3110 | 6 | 000222
                          4 | 7 | 6
                          0 | 8 |
```

Chapter Quick Quiz

1. The class width is 1.00

2. The class boundaries are –0.005 and 0.995

3. No

4. 61 min., 62 min., 62 min., 62 min., 62 min., 67 min., and 69 min.

5. No

6. Bar graph

8. Pareto Chart

7. Scatterplot

9. The distribution of the data set

10. The bars of the histogram start relatively low, increase to a maximum value and then decrease. Also, the histogram is symmetric with the left half being roughly a mirror image of the right half.

Review Exercises

1.

Volume (cm^3)	Frequency
900 – 999	1
1000 – 1099	10
1100 – 1199	4
1200 – 1299	3
1300 – 1399	1
1400 – 1499	1

2. No, the distribution does not appear to be normal because the graph is not symmetric.

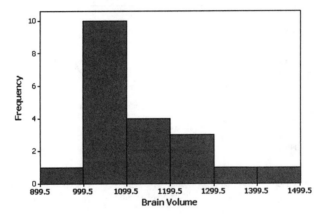

3. Although there are differences among the frequencies of the digits, the differences are not too extreme given the relatively small sample size, so the lottery appears to be fair.

4. The sample size is not large enough to reveal the true nature of the distribution of IQ scores for the population from which the sample is obtained.

```
 8 |7 7 9
 9 |6 6
10 |1 3 3
```

5.   A time-series graph is best. It suggests that the amounts of carbon monoxide emissions in the United States are increasing.

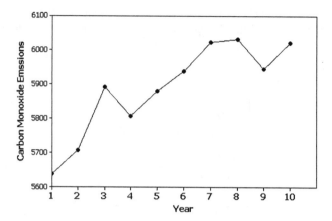

6.   A scatterplot is best. The scatterplot does not suggest that there is a relationship.

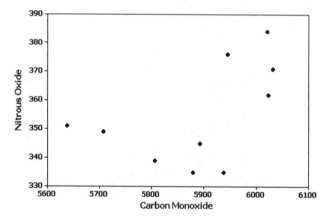

7.   A Pareto chart is best.

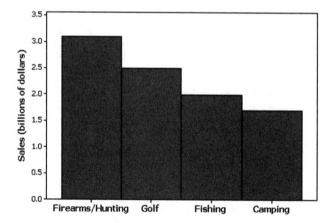

Cumulative Review Exercises

1.   Pareto chart.

2.   Nominal, because the responses consist of names only. The responses do not measure or count anything, and they cannot be arranged in order according to some quantitative scale.

3. Voluntary response sample. The voluntary response sample is not likely to be representative of the population, because those with special interests or strong feelings about the topic are more likely than others to respond and their views might be very different from those of the general population.

4. By using a vertical scale that does not begin at 0, the graph exaggerates the differences in the numbers of responses. The graph could be modified by starting the vertical scale at 0 instead of 50.

5. The percentage is $\frac{241}{641} = 0.376 = 37.6\%$. Because the percentage is based on a sample and not a population that percentage is a statistic.

6.

Grooming Time (min.)	Frequency
0 – 9	2
10 – 19	3
20 – 29	9
30 – 39	4
40 – 49	2

7. Because the frequencies increase to a maximum and then decrease and the left half of the histogram is roughly a mirror image of the right half, the data appear to be from a population with a normal distribution.

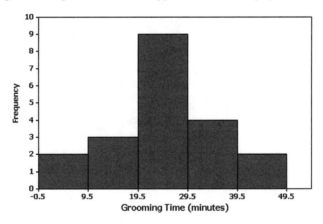

8. Stemplot

```
0 | 0 5
1 | 2 5 5
2 | 0 2 4 5 5 5 7 7 8
3 | 0 0 5 5
4 | 0 5
```

Chapter 3: Statistics for Describing, Exploring, and Comparing Data

Section 3-2

1. No. The numbers do not measure or count anything, so the mean would be a meaningless statistic.

3. No. The price exactly in between the highest and lowest is the midrange, not the median.

5. The mean is $\dfrac{332 + 302 + 235 + 225 + 100 + 90 + 88 + 84 + 75 + 67}{10} = 159.8$ million.

    The median is $\dfrac{90+100}{2} = \$95$ million.

    There is no mode.

    The midrange is $\dfrac{332+67}{2} = \$199.5$ million.

    Apart from the obvious and trivial fact that the mean annual earnings of all celebrities is less than \$332 million, nothing meaningful can be known about the mean of the population.

7. The mean is $\dfrac{371+356+393+544+326+520+501}{7} = 430.1$ hic.

    The median is 393 hic.
    There is no mode.

    The midrange is $\dfrac{326+544}{2} = 435$ hic.

    The safest of these cars appears to be the Hyundai Elantra. Because the measurements appear to vary substantially from a low of 326 hic to a high of 544 hic, it appears that some small cars are considerably safer than others.

9. The mean is $\dfrac{58+22+27+29+21+10+10+8+7+9+11+9+4+4}{14} = \$16.4$ million.

    The median is $\dfrac{10+10}{2} = 10$ million.

    The modes are \$4 million, \$9 million, and \$10 million.

    The midrange is $\dfrac{4+58}{2} = \$31$ million.

    The measures of center do not reveal anything about the pattern of the data over time, and that pattern is a key component of a movie's success. The first amount is highest for the opening day when many Harry Potter fans are most eager to see the movie, the third and fourth values are from the first Friday and the first Saturday, which are the popular weekend days when movie attendance tends to spike.

11. The mean is $\dfrac{55.99+69.99+48.95+48.92+71.77+59.68}{6} = \$59.22$.

    The median is $\dfrac{55.99+59.68}{2} = \$57.84$.

    There is no mode.

    The midrange is $\dfrac{48.92+71.77}{2} = \$60.35$.

    None of the measures of center are most important here. The most relevant statistic in this case is the minimum value of \$48.92, because that is the lowest price for the software. Here, we generally care about the lowest price not the mean price or median price.

13. The mean is $\dfrac{3+6.5+6+5.5+20.5+7.5+12+11.5+17.5}{10}=11.05\mu g/g$.

   The median is $\dfrac{7.5+11.5}{2}=9.5\ \mu g/g$.

   The mode is $20.5\ \mu g/g$.

   The midrange is $\dfrac{3+20.5}{2}=11.75\ \mu g/g$.

   There is not enough information given here to assess the true danger of these drugs, but ingestion of any lead is generally detrimental to good health. All of the decimal values are either 0 or 5, so it appears that the lead concentrations were rounded to the nearest one-half unit of measurement.

15. The mean is $\dfrac{4+4+4+4+4+4+4.5+4.5+4.5+4.5+4.5+4.5+6+6+8+9+9+13+13+15}{20}=$

   6.5 years.

   The median is $\dfrac{4.5+4.5}{2}=4.5$ years.

   The modes are 4 years and 4.5 years.

   The midrange is $\dfrac{4+15}{2}=9.5$ years.

   It is common to earn a bachelor's degree in four years, but the typical college student requires more than four years.

17. The mean is $\dfrac{(-15)+(-18)+(-32)+(-21)+(-9)+(-32)+11+2}{8}=-14.3\,\text{min.}$

   The median is $\dfrac{(-15)+(-18)}{2}=-16.5$.

   The mode is –32 min.

   The midrange is $\dfrac{(-32)+11}{2}=-10.5$.

   Because the measures of center are all negative values, it appears that the flights tend to arrive early before the scheduled arrival times, so the on-time performance appears to be very good.

19. The mean is $\dfrac{9+23+25+88+12+19+74+77+76+73+78}{11}=50.4$.

   The median is 73.
   There is no mode.

   The midrange is $\dfrac{9+78}{2}=48.5$.

   The numbers do not measure or count anything; they are simply replacements for names. The data are at the nominal level of measurement, and it makes no sense to compute the measures of center for these data.

21. White drivers' mean is 73 mi/h.
   White drivers' median is 73 mi/h.
   African American drivers' mean is 74 mi/h.
   African American drivers' median is 74 mi/h.
   Although the African American drivers have a mean speed greater than the white drivers, the difference is very small, so it appears that drivers of both races appear to speed about the same amount.

23. Obama had a mean of $653.9 and a median of $452.
   McCain had a mean of $458.5 and a median of $350.
   The contributions appear to favor Obama because his mean and median are substantially higher. With 66 contributions to Obama and 20 to McCain, Obama collected substantially more in total contributions.

25. The mean is 1.184 the median is 1.235. Yes, it is an outlier because it is a value that is very far away from all the other sample values.

27. The mean is 15 years and the median is 16 years. Presidents receive Secret Service protection after they leave office, so the mean is helpful in planning for the cost and resources used for that protection.

29. $\dfrac{27(24.5)+34(34.5)+13(44.5)+2(54.5)+4(64.5)+1(74.5)+1(84.5)}{27+34+13+2+4+1+1} = 35.8$. This result is quite close to the mean of 35.9 years found by using the original list of data values.

31. $\dfrac{4(54.5)+10(64.5)+25(74.5)+43(84.5)+26(94.5)+8(104.5)+3(114.5)+2(124.5)}{4+10+25+43+26+8+3+2} = 84.7$. This result is close to the mean of 84.4 found using the original list of data values.

33. a. $x = 5(0.62) - 0.3 - 0.4 - 1.1 - 0.7 = 0.6$ parts per million

    b. $n-1$

35. The mean is 39.07, the 10% trimmed mean is 27.677, and the 20% trimmed mean is 27.176. By deleting the outlier of 472.2, the trimmed means are substantially different from the untrimmed mean.

37. The geometric mean is $\sqrt[5]{1.017 \cdot 1.037 \cdot 1.052 \cdot 1.051 \cdot 1.027} = 1.036711036$, or 1.0367 when rounded. Single percentage growth rate is 3.67%. The result is not exactly the same as the mean which is 3.68%.

39. The median is $30 + (10)\left(\dfrac{\dfrac{27+34+13+2+4+1+1+1}{2} - (27+1)}{34} \right) = 33.970588$ years, which is rounded to 34 years. The value of 33 years is better because it is based on the original data and does not involve interpolation.

Section 3-3

1.   The IQ scores of a class of statistics students should have less variation, because those students are a much more homogeneous group with IQ scores that are likely to be closer together.

3.   Variation is a general descriptive term that refers to the amount of dispersion or spread among the data values, but the variance refers specifically to the square of the standard deviation.

5.   The range is $332 - 67 = 265$ million.

The variance is $s^2 = \dfrac{10(350,292) - (2,553,604)^2}{10(9)} = 10548$ square of million dollars.

The standard deviation is $s = \sqrt{10,548} = \$102.703$ million.

Because the data values are 10 highest from the population, nothing meaningful can be known about the standard deviation of the population.

7.   The range is $544 - 326 = 218$ hic.

The variance is $\dfrac{7(1,342,439) - (9,066,121)}{7(6)} = 7879.8$ hic squared.

The standard deviation is $\sqrt{7879.8} = 88.8$ hic.

Although all of the cars are small, the range from 326 hic to 544hic appears to be relatively large, so the head injury measurements are not about the same.

9. The range is $58 - 4 = \$54$ million.

    The variance is $\dfrac{14(6487) - 52441}{14(13)} = 210.9$ square of million dollars.

    The standard deviation is $\sqrt{210.9} = \$14.5$.

    An investor would care about the gross from opening day and the rate of decline after that, but the measures of center and variation are less important.

11. The range is $\$71.77 - \$48.92 = \$22.85$.

    The variance is $\dfrac{6(21,535.3844) - 126,238}{6(5)} = 99.141$ dollars squared.

    The standard deviation is $\sqrt{99.141} = \$9.957$.

    The measures of variation are not very helpful in trying to find the best deal.

13. The range is $20.5 - 3 = 17.5 \mu g/g$.

    The variance is $\dfrac{10(1596.75) - 12,210.25}{10(9)} = 41.75 \left(\mu g/g\right)^2$.

    The standard deviation is $\sqrt{41.75} = 6.46 \mu g/g$.

    If the medicines contained no lead, all of the measures would be $0 \mu g / g$, and the measures of variation would all be 0 as well.

15. The range is $15 - 4 = 11$ years.

    The variance is $\dfrac{20(1078.5) - 16,900}{20(19)} = 12.3$ years^2.

    The standard deviation is $\sqrt{12.3} = 3.5$ years.

    No, because 12 years is within 2 standard deviations of the mean.

17. The range is $11 - (-32) = 43$ min.

    The variance is $\dfrac{8(3244) - 12,996}{8(7)} = 231.4$ min. squared.

    The standard deviation is $\sqrt{231.4} = 15.2$ min.

    The standard deviation can never be negative.

19. The range is $88 - 9 = 79$.

    The variance is $\dfrac{11(38,078) - 306,916}{11(10)} = 1017.7$.

    The standard deviation is $\sqrt{1017.7} = 31.9$.

    The data are at the nominal level of measurement and it makes no sense to compute the measures of variation for these data.

21. The mean of the White drivers is 73 and the standard deviation is 2.906 the coefficient of variation for the White drivers is $\dfrac{2.906}{73} \cdot 100\% = 4\%$. The mean for the African American 74 and the standard deviation is 2.749 the coefficient of variation for the African American drivers is $\dfrac{2.749}{74} \cdot 100\% = 3.7\%$. The variation is about the same.

23. The mean of Obama contributors is \$654 and the standard deviation is \$523 the coefficient of variation is $\frac{\$523}{\$654} \cdot 100\% = 80\%$. The mean of McCain contributors is \$459 and the standard deviation is \$418 the coefficient of variation is $\frac{\$418}{\$459} \cdot 100\% = 90\%$. The variation among Obama contributors is a little less than the variation among the McCain contributors.

25. The range is 2.95, the variance is 0.345, and the standard deviation is 0.587.

27. The range is 36 years, the variance is 94.5 years squared, and the standard deviation is 9.7 years.

29. The standard deviation $\frac{2.95}{4} = 0.738$, which is not substantially different from 0.587

31. The standard deviation $\frac{36}{4} = 9$ years, this is reasonably close to 9.7 years.

33. No. The pulse rate of 99 beats per minute is between the minimum usual value of 54.3 beats per minute and the maximum usual value of 100.7 beats per minute.

35. Yes. The volume of 11.9 oz. is not between the minimum usual value of 11.97 oz. and the maximum usual value of 12.41 oz.

37. $s = \sqrt{\frac{82(84,408.5) - 8,637,721}{82(81)}} = 12.3$ years. This result is not substantially different from the standard deviation of 11.1 years found from the original list of data values.

39. $s = \sqrt{\frac{121(889,106.69) - 104,941,584.81}{121(120)}} = 13.5$. The result is very close to the standard deviation of 13.4 found from the original list of sample values.

41. a. 95%

    b. 68%

43. At least 75% of women have platelet counts within 2 standard deviations of the mean. The minimum is 150 and the maximum is 410.

45. a. $\sigma^2 = \frac{(2 - 4.33)^2 + (3 - 4.33)^2 + (8 - 4.33)^2}{3} = 6.9 \, \text{min}^2$

    b. The nine possible samples of two values are the following: [(2 min, 2 min), (2 min, 3 min), (2 min, 8 min), (3 min, 2 min), (3 min, 3 min), (3 min, 8 min), (8 min, 2 min), (8 min, 3 min), (8 min, 8 min)] and they have the following corresponding variances: [0, 0.707, 18, 0.707, 0, 12.5, 18, 12.5, 0] which have the mean of 6.934.

    c. The population variances of the nine samples above are [0, 0.3535, 9, 0.3535, 0, 6.25, 9, 6.25, 0]

    d. Part (b), because repeated samples result in variances that target the same value (6.9 min.^{2}) as the population variance. Use division by $n-1$.

    e. No. The mean of the sample variances (6.9 min.^{2}) equals the population variance, but the mean of the sample standard deviations (1.9 min.) does not equal the mean of the population standard deviation (2.6 min.)

Section 3-4

1. Madison's height is below the mean. It is 2.28 standard deviations below the mean.

3. The lowest amount is \$5 million, the first quartile Q_1 is \$47 million, the second quartile Q_2 (or median) is \$104 million, the third quartile Q_3 is \$121 million, and the highest gross amount is \$380 million.

5. a.   The difference is $\$3,670,505 - \$4,939,455 = -\$1,268,950$

    b. $\dfrac{\$1,268,950}{\$7,775,948} = 0.16$ standard deviations

    c. $z = -0.16$

    d.   Usual

7. a.   The difference is $\$1 - \$1,449,779 = -\$1,449,778$

    b. $\dfrac{\$1,449,778}{\$527,651} = 2.75$ standard deviation

    c. $z = -2.75$

    d.   Unusual

9. Z scores of –2 and 2. A z score of –2 means a score of $x = -2 \cdot 15 + 100 = 70$. A z score of 2 means a score of $x = 2 \cdot 15 + 100 = 130$

11. Two standard deviations from the mean: $1.240 - 2 \cdot 0.578 = 0.084$ and $1.240 + 2 \cdot 0.578 = 2.396$

13. The tallest man z score is $z = \dfrac{247 - 175}{7} = 10.29$ the tallest women z score is $z = \dfrac{236 - 162}{6} = 12.33$. De-Fen Yao is relatively taller, because her z score of 12.33, which is greater than the z score of 10.29 for Sultan Kosen. De-Fen Yao is more standard deviations above the mean than Sultan Kosen.

15. The SAT score of 1490 has a z score of $z = \dfrac{1490 - 1518}{325} = -0.09$, and the ACT score of 17 has a z score of $z = \dfrac{17 - 21.1}{4.8} = -0.85$. The z score of –0.09 is a larger number than the z score of –0.85, so the SAT score of 1490 is relatively better.

17. The percentile for 213 sec. is $\dfrac{3}{24} \cdot 100 = 13$, so the 13th percentile

19. The percentile for 250 sec. is $\dfrac{12}{24} \cdot 100 = 50$, so the 50th percentile

21. $P_{60} = \dfrac{60 \cdot 24}{100} = 14.4$, pick 15th entry which is 251 sec.

23. $Q_3 = \dfrac{255 + 255}{2} = 255$ sec.

25. $P_{50} = \dfrac{245 + 250}{2} = 247.5$ sec.

27. $P_{25} = Q_1 = 234.5$ sec.

29. The five number summary: 1 sec, 8709 sec, 10,074.5 sec, 11,445 sec, 11,844 sec

31. The five number summary : 4 min, 14 min, 18 min, 32 min, 63 min

33. It appears that males have lower pulse rates than females

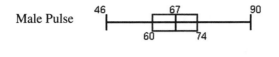

Male Pulse

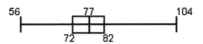

Female Pulse

35. The weights of regular Coke appear to be generally greater than those of diet Coke, probably due to the sugar in cans of regular Coke.

CKREGWT

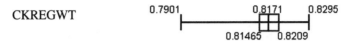

CKDIETWT

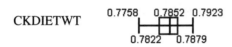

37. Outliers for actresses 60 years, 61 years, 63 years, 70 years, and 80 years. Outliers for actors: 76 years. The modified boxplots show that only one actress has an age that is greater than any actor.

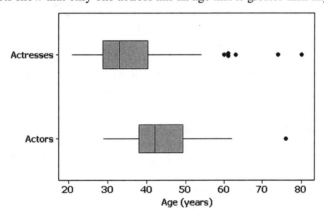

Chapter Quick Quiz

1. The mean is 14 minutes

2. The median is 12 minutes

3. The mode is 12 minutes

4. The variance is $(5 \text{ min})^2 = 25 \text{ min}^2$

5. $z = \dfrac{6-11.4}{7} = -0.77$

6. Standard deviation, variance, range, mean absolute deviation

7. Sample mean $\overline{x}$, population mean μ

8. s, σ, s^2, σ^2

9. 75%

10. Minimum, first quartile Q_1, second quartile Q_2 (or median), third quartile Q_3, maximum

Review Exercises

1. a. $\overline{x} = \dfrac{1550+1642+1538+1497+1571}{5} = 1559.6 \text{ mm}$

   b. The median is 1550 mm

   c. There is no mode

1.   (continued)

   d.   The midrange is $\dfrac{1497+1642}{2}=1569.5\,\text{mm}$

   e.   The range is $1642-1497=145\,\text{mm}$

   f. $s=\sqrt{\dfrac{(1550-1559.6)^2+(1642-1559.6)^2+(1538-1559.6)^2+(1497-1559.6)^2+(1571-1559.6)^2}{5-1}}$

   $=53.37\;\text{mm}$

   g. $s^2=53.37^2=2849.3\,\text{mm}^2$

   h. $Q_1=\dfrac{25\cdot 5}{100}=1.25$, pick second entry (in ordered list) which is 1538 mm

   i. $Q_3=\dfrac{75\cdot 5}{100}=3.75$, pick the fourth entry (in the ordered list) which is 1571 mm

2. $z=\dfrac{1642-1559.6}{53.37}=1.54$. The eye height is not unusual because its z score is between –2 and 2, so it is within two standard deviations of the mean.

3.   The five number summary: 1497, 1538, 1550, 1571, 1642

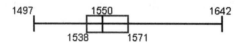

   Because the boxplot shows a distribution of data that is roughly symmetric, the data could be from a population with a normal distribution, but the data are not necessarily from a population with a normal distribution, because there is no way to determine whether a histogram is roughly a bell shape.

4.   The mean is 10053.5. The ZIP codes do not measure or count anything. They are at the nominal level of measurement, so the mean is a meaningless statistic.

5.   The male z score is $z=\dfrac{28-26.601}{5.359}=0.26$. The female z score is $z=\dfrac{29-28.441}{7.394}=0.08$. The male has a larger relative BMI because the male has the larger z score.

6.   a.   The answers may vary but a mean around $8 or $9 is reasonable.

   b.   A reasonable standard deviation would be around $1 or $2.

7.   Based on a minimum age of 23 years and a maximum age of 70 years an estimate of the age standard deviation would be $\dfrac{70-23}{4}=11.75$ years.

8.   A minimum usual sitting height of $914-2\cdot 36=842\,\text{mm}$ and a maximum sitting height of $914+2\cdot 36=986\,\text{mm}$. The maximum usual height of 986 mm is more relevant for designing overhead bin storage.

9.   The minimum value is 963 cm^3, the first quartile is 1034.5 cm^3, the second quartile (or median) is 1079 cm^3, the third quartile is 1188.5 cm^3, and the maximum value is 1439 cm^3.

10. The median would be better because it is not affected much by the one very large income.

Cumulative Review Exercises

1.   a.   Continuous

   b.   Ratio

2.

Hand Length (mm)	Frequency
150 – 159	1
160 – 169	0
170 – 179	2
180 – 189	0
190 – 199	3
200 – 209	1
210 – 219	1

3. Hand length histogram

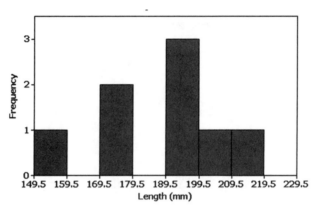

4.

```
15 | 8
16 |
17 | 3 9
18 |
19 | 5 6 9
20 | 7
21 | 4
```

5. a. $\bar{x} = \dfrac{173+179+207+158+196+195+214+199}{8} = 190.1 \text{ mm}$

   b. The median is 195.5 mm

   c. $(173-190.1)^2 + (179-190.1)^2 + (207-190.1)^2 + (158-190.1)^2 + (196-190.1)^2$
   $+ (195-190.1)^2 + (214-190.1)^2 + (199-190.1)^2 = 2440.88$

   $s = \sqrt{\dfrac{2440.88}{7}} = 18.7 \text{ mm mm}$

   d. $s^2 = 18.7^2 = 348.7 \text{ mm}^2$

   e. The range is $214 - 158 = 56 \text{ mm}$

6. Yes. The frequencies increase to a maximum, and then they decrease. Also, the frequencies preceding the maximum are roughly a mirror image of those that follow the maximum.

7. No. Even though the sample is large, it is a voluntary response sample, so the responses cannot be considered to be representative of the population of the United States.

8. The vertical scale does not begin at 0, so the differences among different outcomes are exaggerated.

Chapter 4: Probability

Section 4-2

1. $P(A) = \dfrac{1}{10,000} = 0.0001$, $P(\bar{A}) = 1 - \dfrac{1}{10,000} = \dfrac{9999}{10,000} = 0.9999$

3. Part (c).

5. 5:2, $\dfrac{7}{3}$, −0.9, $\dfrac{456}{123}$

7. $\dfrac{1}{5}$ or 0.2

9. Unlikely, neither unusually low nor unusually high

11. Unlikely, unusually low

13. $\dfrac{1}{4}$ or 0.25

15. $\dfrac{1}{2}$ or 0.5

17. $\dfrac{1}{5}$ or 0.2

19. 0

21. $\dfrac{6}{1000}$ or 0.006. The employer would suffer because it would be at a risk by hiring someone who uses drugs.

23. $\dfrac{50}{1000}$ or 0.05. This result is not close to the probability of 0.134 for a positive test result.

25. $\dfrac{879}{945}$ or 0.93. Yes, the technique appears to be effective.

27. $\dfrac{304}{300,000,000}$ or 0.00000101. No, the probability of being struck is much greater on an open golf course during a thunder storm. The golfer should seek shelter.

29. a. $\dfrac{1}{365}$

    b. Yes

    c. He already knew

    d. 0

31. $\dfrac{10,427,000}{135,933,000}$ or 0.0767. No, a crash is not unlikely. Given that car crashes are so common, we should take precautions such as not driving after drinking and not using a cell phone or texting.

33. $\dfrac{8}{8+804} = 0.00985$. It is unlikely

35. $\dfrac{8}{492+8+306} = 0.00993$. Yes, it is unlikely. The middle seat lacks an outside view, easy access to the aisle, and a passenger in the middle seat has passengers on both sides instead of on one side only.

37. $\dfrac{3}{8}$ or 0.375

39. {bb, bg, gb, gg}; $\dfrac{1}{2}$ or 0.5

41. a. brown /brown, brown/blue, blue/brown, blue/blue

    b. $\dfrac{1}{4}$

    c. $\dfrac{3}{4}$

43. a.   999 : 1

    b.   499 : 1

    c.   The description is not accurate. The
         odds against winning are 999:1 and the
         odds in favor are 1:999, not 1:1000

45. a. $16

    b.   8 : 1

    c.   About 9.75 : 1, which becomes 39 : 4

    d. $21.50

47. Relative risk: $\dfrac{\dfrac{26}{2103}}{\dfrac{22}{1671}} = 0.939$
    Odds ratio: $\dfrac{\dfrac{\dfrac{26}{2103}}{1 - \dfrac{26}{2103}}}{\dfrac{\dfrac{22}{1671}}{1 - \dfrac{22}{1671}}} = 0.938$

The probability of a headache with Nasonex (0.0124) is slightly less than the probability of a headache with the placebo (0.0132), so Nasonex does not appear to pose a risk of headache.

49. $\dfrac{1}{4}$

Section 4-3

1. Based on the rule of the complements, the sum of $P(A)$ and its complement must always be 1, so the sum cannot be 0.5

3. Because it is possible to select someone who is male and a Republican, events M and R are not disjoint. Both events can occur at the same time when someone is randomly selected.

5. Disjoint

7. Not disjoint

9. Disjoint

11. Not disjoint

13. $1 - 0.47 = 0.53$

15. $P(\overline{D}) = 0.45$, where $P(\overline{D})$ is the probability of randomly selecting someone who does not choose a direct in-person encounter as the most fun way to flirt.

17. 1

19. $\dfrac{90 + 860 + 6}{1000} = 0.956$

21. $\dfrac{13}{28}$ or 0.464. That probability is not as high as it should be.

23. $\dfrac{16}{28}$ or 0.571

25. a. $\dfrac{11}{14} = 0.786$ or 78.6%

    b. $\dfrac{2}{14} = 0.143$ or 14.3%

    c.   The physicians given the labels with concentrations appear to have done much better. The results suggest that labels described as concentrations are much better than labels described as ratios.

Use the following table for Exercises 27–31

	Age						Total
	18–21	22–29	30–39	40–49	50–59	60 and over	
Responded	73	255	245	136	138	202	**1049**
Refused	11	20	33	16	27	49	**156**
Total	**84**	**275**	**278**	**152**	**165**	**251**	**1205**

27. $\dfrac{156}{1205} = 0.129$. Yes. A high refusal rate results in a sample that is not necessarily representative of the population, because those who refuse may well constitute a particular group with opinions different from others.

29. $\dfrac{1049}{1205} + \dfrac{84}{1205} - \dfrac{73}{1205} = \dfrac{1060}{1205} = 0.88$

31. $\dfrac{1049}{1205} + \dfrac{275 + 278}{1205} - \dfrac{255 + 245}{1205} = \dfrac{1102}{1205} = 0.915$

33. 300

	Positive Test Result	Negative Test Result	Total
Subject Used Marijuana	119	3	**122**
Subject Did not Use Marijuana	24	154	**178**
Total	**143**	**157**	**300**

35. $\dfrac{3 + 154 + 24}{300} = 0.603$

37. $\dfrac{27}{300} = 0.09$. With an error rate of 0.09 or 9%, the test does not appear to be highly accurate.

39. $\dfrac{3}{4}$ or 0.75

41. $P(A \text{ or } B) = P(A) + P(B) - 2P(A \text{ and } B)$

43. a. $1 - P(A) - P(B) + P(A \text{ and } B)$

   b. $1 - P(A \text{ and } B)$

   c. No

Section 4-4

1. The probability that the second selected senator is a Democrat given that the first selected senator was a Republican.

3. False. The events are dependent because the radio and air conditioner are both powered by the same electrical system. If you find that your car's radio does not work, there is a greater probability that the air conditioner will also not work.

5. a. The events are dependent

   b. $\dfrac{1}{132}$ or 0.00758

7. a. Independent

   b. $\dfrac{1}{12}$ or 0.0833

9. a.   Independent

    b. $\dfrac{5}{222} \cdot \dfrac{5}{222} = 0.000507$

11. a.   Dependent

    b. $\dfrac{58}{100} \cdot \dfrac{1}{99} = 0.00586$

13. a. $\dfrac{90}{1000} \cdot \dfrac{90}{1000} = 0.0081$. Yes, it is unlikely

    b. $\dfrac{90}{1000} \cdot \dfrac{89}{999} = 0.00802$. Yes, it is unlikely

15. a. $\dfrac{904}{1000} \cdot \dfrac{904}{1000} \cdot \dfrac{904}{1000} = 0.739$. No, it is not unlikely

    b. $\dfrac{904}{1000} \cdot \dfrac{903}{999} \cdot \dfrac{902}{998} = 0.739$. No , it is not unlikely

17. $\dfrac{8330}{8834} \cdot \dfrac{8329}{8833} \cdot \dfrac{8328}{8832} = 0.838$. No, the entire batch consists of malfunctioning pacemakers.

19. a. $\dfrac{2}{100} = 0.02$

    b. $\dfrac{2}{100} \cdot \dfrac{2}{100} = 0.0004$

    c. $\dfrac{2}{100} \cdot \dfrac{2}{100} \cdot \dfrac{2}{100} = 0.000008$

    d.   By using one backup drive, the probability of failure is 0.02, and with three independent disk drives, the probability drops to 0.000008. By changing from one drive to three, the likelihood of failure drops from 1 chance in 50 to only 1 chance in 125,000, and that is a very substantial improvement in reliability. BACK UP YOUR DATA.

21. a. $\dfrac{1}{365}$ or 0.00274

    b. $\dfrac{1}{365} \cdot \dfrac{1}{365} = 0.00000751$

    c. $\dfrac{1}{365}$ or 0.00274

23.

	Positive Test Result	Negative Test Result	**Total**
Subject used marijuana	True Positive 119	False Negative 3	**122**
Subject did not use marijuana	False Positive 24	True Negative 154	**178**
Total	**143**	**157**	**300**

$\dfrac{119}{300} \cdot \dfrac{118}{299} + \dfrac{154}{300} \cdot \dfrac{153}{299} + \dfrac{154}{300} \cdot \dfrac{119}{299} + \dfrac{119}{300} \cdot \dfrac{154}{299} = 0.828$. No, it is not unlikely

25. $\dfrac{24}{300} \cdot \dfrac{23}{299} \cdot \dfrac{22}{298} = 0.000454$. Yes, it is unlikely

27. a. $\dfrac{2518-252}{2518}=0.9$

    b. $\left(\dfrac{2518-252}{2518}\right)^{50}=0.00513$. Using the 5% guideline for cumbersome calculations

29. a. $\dfrac{162}{427}\cdot\dfrac{161}{426}=0.143$

    b. $\left(\dfrac{427-162}{427}\right)^{10}=0.00848$. Using the 5% guideline for cumbersome calculations

31. a. $0.99\cdot0.99+0.99\cdot0.01+0.01\cdot0.99=0.9999$

    b. $0.99\cdot0.99=0.9801$

    c. The series arrangement provides better protection.

Section 4-5

1. a. Answers vary, but 0.98 is a reasonable estimate.

   b. Answers vary, but 0.999 is a reasonable estimate.

3. The probability that the polygraph indicates lying given that the subject is actually telling the truth.

5. At least one of the five children is a boy. $\dfrac{31}{32}$ or 0.969

7. None of the digits is 0. $\left(\dfrac{9}{10}\right)^{4}=0.656$

9. $1-\left(\dfrac{4}{5}\right)^{10}=0.893$. The chance of passing is reasonably good

11. 0.5 or 50%   13. $1-(0.512)^{5}=0.965$

15. $1-(1-0.0423)^{3}=0.122$. Given that the three cars are in the same family, they are not randomly selected and there is a good chance that the family members have similar driving habits, so the probability might not be accurate.

17. $1-(1-0.67)^{4}=0.988$. It is very possible that the result is not valid because it is based on data from a voluntary response survey.

19. $\dfrac{90}{950}$ or 0.0947. This is the probability of the test making it appear that the subject uses drugs when the subject is not a drug user.

21. $\dfrac{6}{866}$ or 0.00693. This result is substantially different from the result found in Exercise 20. The probabilities P(subject uses drugs | negative test result) and P(negative test result | subject uses drugs) are not equal.

23. $\dfrac{44}{134}$ or 0.328   25. a. $\dfrac{1}{3}$ or 0.333

                                                      b. $\dfrac{5}{10}$ or 0.5

27. $\dfrac{10}{20}$ or 0.5

29. a. $1-(0.02)^2 = 0.9996$

    b. $1-(0.02)^3 = 0.999992$

31. $1-(1-0.134)^8 = 0.684$. The probability is not low, so further testing of the individual samples will be necessary for about 68% of the combined samples.

33. a. $\dfrac{365}{365} \cdot \dfrac{364}{365} \cdot \dfrac{363}{365} \cdot \ldots \cdot \dfrac{341}{365} = 0.431$

    b. $1-0.431 = 0.569$

35. a. $\dfrac{0.8 \cdot 0.01}{0.8 \cdot 0.01 + 0.1 \cdot 0.99} = 0.0748$

    b. 0.8

    c. The estimate of 75% is dramatically greater than the actual rate of 7.48%. They exhibited confusion of the inverse. A consequence is that they would unnecessarily alarm patients who are benign, and they might start treatments that are not necessary.

Section 4-6

1. The symbol ! is the factorial symbol that represents the product of decreasing whole numbers, as in $4! = 4 \cdot 3 \cdot 2 \cdot 1 = 24$. Four people can stand in line 24 different ways.

3. Because repetition is allowed, numbers are selected with replacement, so neither of the two permutation rules applies. The fundamental counting rule can be used to show that the number of possible outcomes is $10 \cdot 10 \cdot 10 \cdot 10 = 10,000$, so the probability of winning is $\dfrac{1}{10,000}$.

5. $\dfrac{1}{10} \cdot \dfrac{1}{10} \cdot \dfrac{1}{10} \cdot \dfrac{1}{10} = \dfrac{1}{10,000}$

7. $\dfrac{1}{9} \cdot \dfrac{1}{8} \cdot \dfrac{1}{7} \cdot \dfrac{1}{6} \cdot \dfrac{1}{5} \cdot \dfrac{1}{4} \cdot \dfrac{1}{3} \cdot \dfrac{1}{2} \cdot \dfrac{1}{1} = \dfrac{1}{9!} = \dfrac{1}{362,880}$

9. The number of combinations is $\dfrac{27!}{(27-12)!12!} = 17,383,860$. Because that number is so large, it is not practical to make a different CD for each possible combination.

11. $\dfrac{1}{50} \cdot \dfrac{1}{49} \cdot \dfrac{1}{48} \cdot \dfrac{1}{47} = \dfrac{1}{5,527,200}$. No 5,527,200 is too many possibilities to list.

13. $\dfrac{11!}{4!4!2!} = 34,650$

15. $\dfrac{44!}{(44-6)!6!} = 7,059,052$. The probability is $\dfrac{1}{7,059,052}$

17. $\dfrac{1}{4!} = \dfrac{1}{24}$

19. a. $\dfrac{41!}{(41-5)!5!}=749{,}398$.

      The probability is $\dfrac{1}{749{,}398}$

   b. $\dfrac{1}{10^{4}}=\dfrac{1}{10{,}000}$

   c.   \$10,000

21. a. $\dfrac{12!}{(12-4)!}=11{,}880$

   b. $\dfrac{12!}{(12-4)!4!}=495$

   c. $\dfrac{1}{495}$

23. The number of possible "combinations" is $50\cdot50\cdot50=125{,}000$. The fundamental counting rule can be used. The different possible codes are ordered sequences of numbers, not combinations, so the name of "combination lock" is not appropriate. Given that "fundamental counting rule lock" is a bit awkward, a better name would be something like "number lock".

25. $5!=120$; AMITY; $\dfrac{1}{120}$

27. $\dfrac{5!}{5!0!}+\dfrac{5!}{4!1!}+\dfrac{5!}{3!2!}+\dfrac{5!}{2!3!}=26$

29. $\dfrac{5!}{(5-2)!2!}=10$

31. $4\cdot4\cdot4=64$

33. $\dfrac{1}{{}_{59}C_5\cdot{}_{39}C_1}=\dfrac{1}{195{,}249{,}054}$

35. $\dfrac{2}{{}_{10}C_5}=\dfrac{2}{252}$. Yes, if everyone treated is of one gender while everyone in the placebo group is of the opposite gender, you would not know if different reactions are due to the treatment or gender.

37. $26+26\cdot36+26\cdot36^{2}+26\cdot36^{3}+26\cdot36^{4}+26\cdot36^{5}+26\cdot36^{6}+26\cdot36^{7}=2{,}095{,}681{,}645{,}538$

39. 12 ways: {25p, 1n 20p, 2n 15p, 3n 10p, 4n 5p, 5n, 1d 15p, 1d 1n 10p, 1d 2n 5p, 1d 3n, 2d 5p,2d 1n} (Note: 25p represents 25 pennies, etc.)

Chapter Quick Quiz

1. 0 (not an option)

2. $\dfrac{10-3}{10}=0.7$

3. 1 (all days contain the letter y)

4. $0.2\cdot0.2=0.04$

5. Answers vary, but an answer such as 0.01 or lower is reasonable

6. $\dfrac{288+224}{201+126+288+224}=\dfrac{512}{839}=0.61$

7. $\dfrac{224+288+201}{839}=\dfrac{713}{839}=0.85$

8. $\dfrac{126}{839}=0.15$

9. $\dfrac{126}{839}\cdot\dfrac{125}{839}=0.0224$

10. $\dfrac{126}{126+224}=\dfrac{126}{350}=0.36$

Review Exercises

1. $\dfrac{392+58}{1028}=0.438$

2. $\dfrac{392}{392+564}=0.41$

3. $\dfrac{58}{58+14}=0.806$

4. It appears that you have a substantially better chance of avoiding prison if you enter a guilty plea.

5. $\dfrac{392+58}{1028}+\dfrac{392+564}{1028}-\dfrac{392}{1028}=0.986$

8. $\dfrac{72}{1028}+\dfrac{578}{1028}-\dfrac{14}{1028}=0.619$

6. $\dfrac{450}{1028}\cdot\dfrac{449}{1027}=0.191$

9. $\dfrac{392}{1028}=0.381$

7. $\dfrac{72}{1028}\cdot\dfrac{71}{1027}=0.00484$

10. $\dfrac{14}{1028}=0.0136$

11. Answers vary, but DuPont data show that about 8% of cars are red, so any estimate between 0.01 and 0.2 would be reasonable.

12. a. $1-0.35=.65$

    b. $(0.35)^4=0.015$

    c.   Yes, because the probability is so small

13. a. $\dfrac{1}{365}$

15. $\dfrac{1}{{}_{42}C_6}=\dfrac{1}{5,245,786}$

    b. $\dfrac{31}{365}$

16. $\dfrac{1}{{}_{39}C_5}=\dfrac{1}{575,757}$

    c.   Answers vary, but it is probably small, such as 0.02

    d.   Yes

17. $\dfrac{1}{10}\cdot\dfrac{1}{10}\cdot\dfrac{1}{10}=\dfrac{1}{1000}$

14. $1-\left(1-\dfrac{213}{100,000}\right)^{10}=0.0211$. No

18. ${}_{12}P_3=1320$. The probability is $\dfrac{1}{1320}$

Cumulative Review Exercises

1. a.   The mean of –8.9 years is not close to the value of 0 years that would be expected with no gender discrepancy.

   b.   The median of –13.5 years is not close to the value of 0 years that would be expected with no gender discrepancy.

   c. $s=\sqrt{\dfrac{\left(-20-(-8.9)\right)^2+\left(-15-(-8.9)\right)^2+...+\left(-15-(-8.9)\right)^2}{11}}=10.6\,\text{years}.$

   d. $s^2=(10.6)^2=113.2\,\text{years}^2$

   e. $Q_1=-15\,\text{years}$

   f. $Q_3=-5\,\text{years}$

   g.   The boxplot suggests that the data have a distribution that is skewed.

2. a. $z = \dfrac{100-77.5}{11.6} = 1.94$. No, the pulse rate of 100 beats per minute is within 2 standard deviations away from the mean, so it is not unusual.

   b. $z = \dfrac{50-77.5}{11.6} = -2.37$. Yes, the pulse rate of 50 beats per minutes is more than 2 standard deviations away from the mean so it is unusual.

   c. Yes, because the probability of $\dfrac{1}{256}$ (or 0.0039) is so small.

   d. No, because the probability of $\dfrac{1}{8}$ (or 0.125) is not very small.

3. a. $\dfrac{2346}{5100} = .46 = 46\%$

   b. $0.46 = 46\%$

   c. Stratified sample

4. The graph is misleading because the vertical scale does not start at 0. The vertical scale starts at the frequency of 500 instead of 0, so the difference between the two response rates is exaggerated. The graph incorrectly makes it appear that "no" responses occurred 60 times more often than the number of "yes" responses, but comparisons of the actual frequencies shows that the "no" responses occurred about four times more often than the number of "yes" responses.

5. a. A convenience sample

   b. If the students at the college are mostly from a surrounding region that includes a large proportion of one ethnic group, the results will not reflect the general population of the United States.

   c. $0.35 + 0.4 = 0.75$

   d. $1 - (0.6)^2 = 0.64$

6. The straight-line pattern of the points suggests that there is a correlation between chest size and weight.

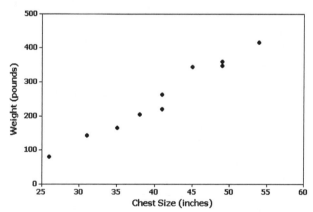

7. a. $\dfrac{1}{{}_{39}C_5} = \dfrac{1}{575,757}$

   b. $\dfrac{1}{19}$

   c. $\dfrac{1}{{}_{39}C_5 \cdot {}_{19}C_1} = \dfrac{1}{10,939,383}$

Chapter 5: Discrete Probability Distributions

Section 5-2

1.   The random variable is x, which is the number of girls in three births. The possible values of x are 0, 1, 2, and 3. The values of the random variable x are numerical.

3.   Table 5-7 does describe a probability distribution because the three requirements are satisfied. First, the variable x is a numerical random variable and its values are associated with probabilities. Second, $\Sigma P(x) = 0.125 + 0.375 + 0.375 + 0.125 = 1$ as required.   Third, each of the probabilities is between 0 and 1 inclusive, as required.

5.   a.   Continuous random variable
     b.   Discrete random variable
     c.   Not a random variable

     d.   Discrete random variable
     e.   Continuous random variable
     f.   Discrete random variable

7.   Probability distribution with
$$\mu = (0 \cdot 0.0625) + (1 \cdot 0.25) + (2 \cdot 0.375) + (3 \cdot 0.25) + (4 \cdot 0.0625) = 2$$
$$\sigma = \sqrt{(0-2)^2 \cdot 0.0625 + (1-2)^2 \cdot 0.25 + (2-2)^2 \cdot 0.375 + (3-2)^2 \cdot 0.25 + (4-2)^2 \cdot 0.0625} = 1$$

9.   Not a probability distribution because the sum of the probabilities is 0.601, which is not 1 as required. Also, Ted clearly needs a new approach.

11. Probability distribution with
$$\mu = (0 \cdot 0.041) + (1 \cdot 0.2) + (2 \cdot 0.367) + (3 \cdot 0.299) + (4 \cdot 0.092) = 2.2$$
$$\sigma = \sqrt{(0-2.2)^2 \cdot 0.041 + (1-2.2)^2 \cdot 0.2 + (2-2.2)^2 \cdot 0.367 + (3-2.2)^2 \cdot 0.299 + (4-2.2)^2 \cdot 0.092} = 1$$

13. Not a probability distribution because the responses are not values of a numerical random variable. Also, sum of the probabilities is 1.18 instead of 1 as required.

15. $$\mu = (0 \cdot 0.001) + (1 \cdot 0.01) + (2 \cdot 0.044) + \ldots + (9 \cdot 0.01) + (10 \cdot 0.001) = 5$$
$$\sigma = \sqrt{(0-5)^2 \cdot 0.001 + (1-5)^2 \cdot 0.01 + (2-5)^2 \cdot 0.044 + \ldots + (9-5)^2 \cdot 0.01 + (10-5)^2 \cdot 0.001} = 1.6$$

17. a. $P(X = 8) = 0.044$

     b. $P(X \geq 8) = 0.044 + 0.01 + 0.001 = 0.055$

     c.   The result from part (b)

     d.   No, because the probability of 8 or more girls is 0.055, which is not very low (less than or equal to 0.05)

19. $$\mu = (0 \cdot 0.377) + (1 \cdot 0.399) + (2 \cdot 0.176) + (3 \cdot 0.041) + (4 \cdot 0.005) + (5 \cdot 0) + (6 \cdot 0) = 0.9$$
$$\sigma = \sqrt{(0-0.9)^2 \cdot 0.377 + (1-0.9)^2 \cdot 0.399 + \ldots + (4-0.9)^2 \cdot 0.005 + (5-0.9)^2 \cdot 0 + (6-0.9)^2 \cdot 0}$$
$$= 0.9$$

21. a. $P(X = 3) = 0.041$

     b. $P(X \geq 3) = 0.041 + 0.005 + 0 + 0 = 0.046$

     c.   The probability from part (b)

     d.   Yes, because the probability of three or more failures is 0.046 which is very low (less than or equal to 0.05)

23. a. $10 \cdot 10 \cdot 10 = 1000$

    b. $\dfrac{1}{1000}$

    c. $\$500 - \$1 = \$499$

    d. $-\$1 \cdot 1 + \$500 \cdot \dfrac{1}{1000} = -\$0.50 = -50\,\text{cents}$

    e.   The \$1 bet on the pass line in craps is better because its expected value of –1.4 cents is much greater than the expected value of –50 cents for the Texas Pick 3 lottery.

25. a. $-\$0.26 + \$30 \cdot \dfrac{5}{38} - \$5 \cdot \dfrac{33}{38} = -0.39$

    b.   The bet on the number 27 is better because its expected value of –26 cents is greater than the expected value of –39 cents for the other bet.

Section 5-3

1. The given calculation assumes that the first two adults include Wal-Mart and the last three adults do not include Wal-Mart, but there are other arrangements consisting of two adults who include Wal-Mart and three who do not. The probabilities corresponding to those other arrangements should also be included in the result.

3. Because the 30 selections are made without replacement, they are dependent, not independent. Based on the 5% guideline for cumbersome calculations, the 30 selections can be treated as being independent. (The 30 selections constitute 3% of the population of 1000 responses, and 3% is not more than 5% of the population.) The probability can be found by using the binomial probability formula.

5. Not binomial. Each of the weights has more than two possible outcomes.

7. Binomial

9. Not binomial. Because the senators are selected without replacement, the selections are not independent. (The 5% guideline for cumbersome calculations cannot be applied because the 40 selected senators constitute 40% of the population of 100 senators, and that exceeds 5%.)

11. Binomial. Although the events are not independent, they can be treated as being independent by applying the 5% guideline. The sample size of 380 is no more than 5% of the population of all smartphone users.

13. a. $\dfrac{4}{5} \cdot \dfrac{4}{5} \cdot \dfrac{1}{5} = 0.128$

    b. $\{WWC,\ WCW,\ CWW\}$; 0.128 for each

    c. $0.128 \cdot 3 = 0.384$

15. $_5C_3 \cdot 0.2^3 \cdot 0.8^2 = 0.051$

17. $_5C_3 \cdot 0.2^3 \cdot 0.8^2 + {}_5C_4 \cdot 0.2^4 \cdot 0.8^1 + {}_5C_5 \cdot 0.2^5 \cdot 0.8^0 = 0.057$

19. $_5C_5 \cdot 0.2^0 \cdot 0.8^5 = 0.328$ 23. $_{20}C_{16} \cdot 0.45^{16} \cdot 0.55^4 = 0.00125$

21. $_8C_3 \cdot 0.45^3 \cdot 0.55^5 = 0.257$

25. $P(X \geq 2) = 0.033 + 0.132 + 0.297 + 0.356 + 0.178 = 0.996$; yes

27. $P(X \leq 2) = 0.000 + 0.004 + 0.033 = 0.037$; yes, because the probability of 2 or fewer peas with green pods is small (less than or equal to 0.05).

29. a. $_6C_5 \cdot 0.20^5 \cdot 0.80^1 = 0.002$ (Tech: 0.00154)

   b. $_6C_6 \cdot 0.20^6 \cdot 0.80^0 = 0+ 0+$ (Tech: 0.000064)

   c. $0.002 + 0 = 0.002$ (Tech: 0.00160)

   d.   Yes, the small probability from part (c) suggests that 5 is an unusually high number.

31. a. $_5C_0 \cdot 0.20^0 \cdot 0.80^5 = 0.328$

   b. $_5C_1 \cdot 0.20^1 \cdot 0.80^4 = 0.410\,0.410$

   c. $0.328 + 0.410 = 0.738$ (Tech: 0.737)

   d.   No, the probability from part (c) is not small, so 1 is not an unusually low number

33. $_{20}C_{12} \cdot 0.48^{12} \cdot 0.52^8 = 0.101$. No, because the probability of exactly 12 is 0.101, the probability of 12 or more is greater than 0.101, so the probability of getting 12 or more is not very small, so 12 us not unusually high

35. $_{12}C_{10} \cdot 0.805^{10} \cdot 0.195^2 = 0.287$. No, because the flights all originate from New York, they are not randomly selected flights, so the 80.5% on-time rate might not apply

37. a. $_{12}C_0 \cdot 0.45^0 \cdot 0.55^{12} = 0.000766$

   b. $1 - 0.000766 = 0.999$

   c. $_{12}C_0 \cdot 0.45^0 \cdot 0.55^{12} + _{12}C_1 \cdot 0.45^1 \cdot 0.55^{11} = 0.00829$

   d.   Yes, the very low probability of 0.00829 would suggest that the 45 share value is wrong

39. a. $_{14}C_{13} \cdot 0.5^{13} \cdot 0.5^1 = 0.000854$

   b. $_{14}C_{14} \cdot 0.5^{14} \cdot 0.5^0 = 0.000061$

   c. $_{14}C_{14} \cdot 0.5^{14} \cdot 0.5^0 + _{14}C_{13} \cdot 0.5^{13} \cdot 0.5^1 = 0.000916$

   d.   Yes. The probability of getting 13 girls or a result of 14 girls is 0.000916, so chance does not appear to be a reasonable explanation for the 13 girls. Because 13 is an unusually high number of girls, it appears that the probability of a girl is higher with the XSORT method, and it appears that the XSORT method is effective.

41. $1 - _{24}C_0 \cdot 0.006^0 \cdot 0.994^{24} = 0.134$. It is not unlikely for such a combined sample to test positive.

43. $_{40}C_1 \cdot 0.03^1 \cdot 0.97^{39} + _{40}C_0 \cdot 0.03^0 \cdot 0.97^{40} = 0.662$. The probability shows that about 2/3 of all shipments will be accepted. With about 1/3 of the shipments rejected, the supplier would be wise to improve quality.

45. $P(X = 5) = 0.06(1 - 0.06)^4 = 0.0468$

47. a. $P(4) = \dfrac{6!}{(6-4)!4!} \cdot \dfrac{43!}{(43-6+4)!(6-4)!} \div \dfrac{(6+43)!}{(6+43-6)!6!} = 0.000969$

   b. $P(6) = \dfrac{6!}{(6-6)!6!} \cdot \dfrac{43!}{(43-6+6)!(6-6)!} \div \dfrac{(6+43)!}{(6+43-6)!6!} = 0.0000000715$

   c. $P(0) = \dfrac{6!}{(6-0)!0!} \cdot \dfrac{43!}{(43-6+0)!(6-0)!} \div \dfrac{(6+43)!}{(6+43-6)!6!} = 0.436$

Section 5-4

1. $n = 270, p = 0.07, q = 0.93$

3.   Variance is $150 \cdot 0.933 \cdot 0.067 = 9.4$ executives^2

5. $\mu = np = 60 \cdot 0.2 = 12$ correct guesses and $\sigma = \sqrt{np(1-p)} = \sqrt{60 \cdot 0.2 \cdot 0.8} = 3.1$ correct guesses.
Minimum: $12 - 2(3.1) = 5.8$ correct guesses, maximum: $12 + 2(3.1) = 18.2$ correct guesses.

7. $\mu = np = 1013 \cdot 0.66 = 668.6$ worriers and $\sigma = \sqrt{np(1-p)} = \sqrt{1013 \cdot 0.66 \cdot 0.34} = 15.1$ worriers. Minimum:
$668.6 - 2(15.1) = 638.4$ worriers, maximum: $668.6 + 2(15.1) = 698.8$ worriers

9.   a. $\mu = np = 291 \cdot 0.5 = 145.5$ boys and $\sigma = \sqrt{np(1-p)} = \sqrt{291 \cdot 0.5 \cdot 0.5} = 8.5$ boys

     b.   Yes. Using the range rule of thumb, the minimum value is $145.5 - 2(8.5) = 128.5$ boys and the
          maximum value is $145.5 + 2(8.5) = 162.5$ boys. Because 239 boys is above the range of usual values,
          it is unusually high. Because 239 boys is unusually high, it does appear that the YSORT method of
          gender selection is effective.

11. a. $\mu = np = 100 \cdot 0.20 = 20$ and $\sigma = \sqrt{np(1-p)} = \sqrt{100 \cdot 0.2 \cdot 0.8} = 4$

     b.   No, because 25 orange M&Ms is within the range of usual values of
          $20 - 2(4) = 12$ and $20 + 2(4) = 28$. The claimed rate of 20% does not necessarily appear to be wrong,
          because that rate will usually result in 12 to 28 orange M&Ms (among 100), and the observed number
          of orange M&Ms is within that range.

13. a. $\mu = np = 420,095 \cdot 0.00034 = 142.8$ and $\sigma = \sqrt{np(1-p)} = \sqrt{420,095 \cdot 0.00034 \cdot 0.999666} = 11.9$

     b.   No, 135 is not unusually low or high because it is within the range of usual values
          $142.8 - 2(11.9) = 119$ and $142.8 + 2(11.9) = 166.6$

     c.   Based on the given results, cell phones do not pose a health hazard that increases the likelihood of
          cancer of the brain or nervous system.

15. a. $\mu = np = 2600 \cdot 0.06 = 156$ and $\sigma = \sqrt{np(1-p)} = \sqrt{2600 \cdot 0.06 \cdot 0.94} = 12.1$

     b.   The minimum usual frequency is $156 - 2(12.1) = 131.8$ and the maximum is $156 + 2(12.1) = 180.2$.
          The occurrence of r 178 times is not unusually low or high because it is within the range of usual
          values.

17. a. $\mu = np = 370 \cdot 0.2 = 74$ and $\sigma = \sqrt{np(1-p)} = \sqrt{370 \cdot 0.2 \cdot 0.8} = 7.7$

     b.   The minimum usual number is $74 - 2(7.7) = 58.6$ and the maximum is $74 + 2(7.7) = 89.4$. The value
          of 90 is unusually high because it is above the range of usual values.

19. a. $\mu = np = 30 \cdot \dfrac{1}{365} = 0.0821918$ and $\sigma = \sqrt{np(1-p)} = \sqrt{30 \cdot \dfrac{1}{365} \cdot \dfrac{364}{365}} = 0.2862981$

     b.   The minimum usual value is $0.0821918 - 2(0.2862981) = -0.4904044$ and the maximum is
          $0.0821918 + 2(0.2862981) = 0.654788$. The result of 2 students born on the 4th of July would be
          unusually high, because 2 is above the range of usual values.

21. From the range of usual values we get $\mu = 60$ and $\sigma = 6$. Using the formulas for the mean and the

     standard deviation we get $p = 1 - \dfrac{\sigma^2}{\mu} = 0.4$ which leads to $n = \dfrac{\mu}{p} = 150$ and $q = 0.6$

23. The probability of selecting a girl out of 40 is given by $P(X = n) = \dfrac{_{10}C_n \cdot {}_{30}C_{12-n}}{_{40}C_{12}}$ the following table list

the probabilities of selecting the number of girls from 0 to 10

Number of girls $(X = n)$	Probability $P(X = n)$
0	0.0154815
1	0.0977782
2	0.2420012
3	0.3073032
4	0.2200011
5	0.0918266
6	0.0223190
7	0.0030608
8	0.0002207
9	0.0000073
10	0.0000001

The mean is $\mu = \sum [x \cdot P(x)]$

$\mu = 0 \cdot 0.0154815 + 1 \cdot 0.0977782 + 2 \cdot 0.2420012 + 3 \cdot 0.3073032 + 4 \cdot 0.2200011 + 5 \cdot 0.0918266 +$

$\quad 6 \cdot 0.0223190 + 7 \cdot 0.0030608 + 8 \cdot 0.0002207 + 9 \cdot 0.0000073 + 10 \cdot 0.0000001 = 3$

The standard deviation is $\sigma = \sqrt{\sum [x^2 \cdot P(x)] - \mu^2}$

$\sigma = \sqrt{\begin{array}{l} 0^2 \cdot 0.0154815 + 1^2 \cdot 0.0977782 + 2^2 \cdot 0.2420012 + 3^2 \cdot 0.3073032 + 4^2 \cdot 0.2200011 + 5^2 \cdot 0.0918266 + \\ 6^2 \cdot 0.0223190 + 7^2 \cdot 0.0030608 + 8^2 \cdot 0.0002207 + 9^2 \cdot 0.0000073 + 10^2 \cdot 0.0000001 - 3^3 \end{array}}$

$\quad = 1.3$

Section 5-5

1. $\mu = \dfrac{535}{576} = 0.929$, which is the mean number of hits per region. $x = 2$, because we want the probability

that a randomly selected region had exactly 2 hits, and $e = 2.71828$ which is a constant used in all applications of Formula 5 – 9.

3. With $n = 50$, the first requirement of $n \geq 100$ is not satisfied. With $n = 50$ and $p = 0.001$ the second requirement of $np \leq 10$ is satisfied. Because both requirements are not satisfied, we should not use the Poisson distribution as an approximation to the binomial.

5. $P(0) = \dfrac{8.5^0 \cdot e^{-8.5}}{0!} = 0.000203$; Yes it is unlikely.

7. $P(10) = \dfrac{8.5^{10} \cdot e^{-8.5}}{10!} = 0.11$; No it is not unlikely

9. a. $\mu = \dfrac{268}{41} = 6.5$

   b. $1 - \dfrac{6.5^0 \cdot e^{-6.5}}{0!} = 0.998$

   c. Yes. Based on the result in part (b), we are quite sure (with probability 0.998) that there is at least one earthquake measuring 6.0 or higher on the Richter scale, so there is a very low probability (0.002) that there will be no such earthquake in a year.

11. a. $\mu = \dfrac{22713}{365} = 62.2$

  b. $P(50) = \dfrac{62.2^{50} \cdot e^{62.2}}{50!} = 0.0155$

13. a. $P(2) = \dfrac{0.929^2 \cdot e^{-0.929}}{2!} = 0.17$

  b.   The expected number of regions with exactly 2 hits is 98.2

  c.   The expected number of regions with 2 hits is close to 93, which is the actual number of regions with 2 hits.

15. a. $P(26) = \dfrac{30.4^{26} \cdot e^{-30.4}}{26!} = 0.0558$. The expected value is $34 \cdot 0.0558 = 1.9$ cookies. The expected number of cookies is very close to the actual number of cookies with 26 chocolate chips which is 2.

  b. $P(30) = \dfrac{30.4^{30} \cdot e^{-30.4}}{30!} = 0.0724$. The expected value is $34 \cdot 0.0724 = 2.5$ cookies. The expected number of cookies is very different from the actual number of cookies with 26 chocolate chips which is 6.

17. a.   No. With $n = 12$ and $p = \frac{1}{6}$ the requirement of $n \ge 100$ is not satisfied, so the Poisson distribution is not a good approximation to the binomial distribution.

  b.   No. The Poisson distribution approximation to the binomial distribution yields

$P(3) = \dfrac{2^3 \cdot e^{-2}}{3!} = 0.18$ and the binomial distribution yields $P(3) = {}_{12}C_3 \cdot \left(\dfrac{1}{6}\right)^3 \cdot \left(\dfrac{5}{6}\right)^9 = 0.197$. The Poisson approximation of 0.18 is too far from the correct result of 0.197.

Chapter Quick Quiz

1. Yes

2. $100 \cdot \dfrac{1}{5} = 20$

3. $\sigma = \sqrt{100 \cdot 0.2 \cdot 0.8} = 4$

4. The range of usual values has a minimum value of $200 - 2 \cdot 10 = 180$ and a maximum value of $200 + 2 \cdot 10 = 220$. Therefore, 232 girls in 400 is an unusually high number of girls since it is outside the range of usual values.

5. The range of usual values has a minimum value of $200 - 2 \cdot 10 = 180$ and a maximum value of $200 + 2 \cdot 10 = 220$. Therefore, 185 girls in 400 is not an unusually high number of girls since it is inside the range of usual values.

6. Yes. The sum of the probabilities is 0.999 and it can be considered to be 1.

7. 0+ indicates that the probability is a very small positive number. It does not indicate that it is impossible for none of the five flights to arrive on time.

8. $P(x \ge 3) = 0.198 + 0.409 + 0.338 = 0.945$

9. $\mu = 0 \cdot 0 + 1 \cdot 0.006 + 2 \cdot 0.048 + 3 \cdot 0.198 + 4 \cdot 0.409 + 5 \cdot 0.338 = 4.022$ and

$\sigma = \sqrt{0^2 \cdot 0 + 1^2 \cdot 0.006 + 2^2 \cdot 0.048 + 3^2 \cdot 0.198 + 4^2 \cdot 0.409 + 5^2 \cdot 0.338 - 4.022^2} = 0.893$

The range of usual values is from 2.236 to 5.808. Since zero is outside the range of usual values it is an unusually low number.

10. $\mu = 0 \cdot 0 + 1 \cdot 0.006 + 2 \cdot 0.048 + 3 \cdot 0.198 + 4 \cdot 0.409 + 5 \cdot 0.338 = 4.022$ and

$\sigma = \sqrt{0^2 \cdot 0 + 1^2 \cdot 0.006 + 2^2 \cdot 0.048 + 3^2 \cdot 0.198 + 4^2 \cdot 0.409 + 5^2 \cdot 0.338 - 4.022^2} = 0.893$

The range of usual values is from 2.236 to 5.808. Since 5 is inside the range of usual values it is not an unusually high number.

Review Exercise

1. $P(X = 0) = {_6}C_0 \cdot 0.4^0 \cdot 0.6^6 = 0.0467$ 2. $P(X = 4) = {_6}C_4 \cdot 0.4^4 \cdot 0.6^2 = 0.138$

3. $\mu = 600 \cdot 0.4 = 240$ and $\sigma = \sqrt{600 \cdot 0.4 \cdot 0.6} = 12$. The range of usual values has a minimum of $240 - 2 \cdot 12 = 216$ and a maximum value of $240 + 2 \cdot 12 = 264$. The result of 200 with brown eyes is unusually low.

4. The probability of 239 or fewer (0.484) is relevant for determining whether 239 is an unusually low number. Because that probability is not very small, it appears that 239 is not an unusually low number of people with brown eyes.

5. Yes. The three requirements are satisfied. There is a numerical random variable x and its values are associated with corresponding probabilities. The sum of the probabilities is 1.001, so the sum is 1 when we allow for a small discrepancy due to rounding. Also, each of the probability values is between 0 and 1 inclusive.

6. $\mu = 0 \cdot 0.674 + 1 \cdot 0.28 + 2 \cdot 0.044 + 3 \cdot 0.003 + 4 \cdot 0 = 0.4$ and

$\sigma = \sqrt{0^2 \cdot 0.674 + 1^2 \cdot 0.28 + 2^2 \cdot 0.044 + 3^2 \cdot 0.003 + 4^2 \cdot 0 - 0.4^2} = 0.6$

The range of usual values has a minimum of $0.4 - 2 \cdot 0.6 = -0.8$ and a maximum value of $0.4 + 2 \cdot 0.6 = 1.6$. Yes, 3 is an unusually high number of males with tinnitus among four randomly selected males.

7. The sum of the probabilities is 0.902 which is not 1 as required. Because the three requirements are not satisfied, the given information does not describe a probability distribution.

8. $\$75 \cdot \dfrac{1}{5} + \$300 \cdot \dfrac{1}{5} + \$75,000 \cdot \dfrac{1}{5} + \$500,000 \cdot \dfrac{1}{5} + \$1,000,000 \cdot \dfrac{1}{5} = \$315,075$. Because the offer is well below her expected value, she should continue the game (although the guaranteed prize of $193000 had considerable appeal).

9. a. $\$1,000,000 \cdot \dfrac{1}{900,000,000} + \$100,000 \cdot \dfrac{1}{110,000,000} + \$25,000 \cdot \dfrac{1}{110,000,000}$

    $+ \$5000 \cdot \dfrac{1}{36,667,000} + \$2500 \cdot \dfrac{1}{27,500,000} = \$0.012$

   b. $0.012 minus the cost of the postage stamp. Since the expected value of winning is much smaller than the cost of a postage stamp, it is not worth entering the contest.

10. a. $\mu = \dfrac{18}{30} = 0.6$

    b. $P(0) = \dfrac{0.6^0 \cdot e^{-0.6}}{0!} = 0.549$

    c. $30 \cdot 0.549 = 16.5$ days

    d. The expected number of days is 16.5, and that is reasonably close to the actual number of days which is 18.

Cumulative Review Exercises

1. a.   The mean is $\bar{x} = \dfrac{22.2 + 24.8 + 24.2 + 26.9 + 23.8}{5} = 24.4\,\text{hours}$

   b.   The median is 24.2 hours

   c.   The range is $26.9 - 22.2 = 4.7\,\text{hours}$

   d.   The standard deviation is

   $$s = \sqrt{\dfrac{(22.2 - 24.4)^2 + (24.8 - 24.4)^2 + (24.2 - 24.4)^2 + (26.9 - 24.4)^2 + (23.8 - 24.4)^2}{5}} = 1.7$$

   e.   The variance is 2.9 hours^2

   f.   The minimum is $24.4 - 2\cdot 1.7 = 21\,\text{hours}$ and the maximum is $24.4 + 2\cdot 1.7 = 27.8\,\text{hours}.$

   g.   No, because none of the times are outside the range of the usual values

   h.   Ratio

   i.   Continuous

   j.   The given times come from countries with very different population sizes, so it does not make sense to treat the given times equally. Calculations of statistics should take the different population sizes into account. Also, the sample is very small, and there is no indication that the sample is random.

2. a. $\dfrac{1}{10}\cdot\dfrac{1}{10}\cdot\dfrac{1}{10}\cdot\dfrac{1}{10} = \dfrac{1}{10{,}000} = 0.0001$

   c. $365\cdot 0.0001 = 0.0365$

   d. $P(1) = \dfrac{0.0365^1 \cdot e^{-0.0365}}{1!} = 0.0352$

   b.

x	$P(x)$
–$1	0.9999
$4999	0.0001

   e. $-\$1\cdot 0.9999 + \$4999\cdot 0.0001 = -\$0.50\,\text{or}\,-50\,\text{cents.}$

3. a. $\dfrac{121 + 51}{611} = 0.282$

   e. $\dfrac{121 + 51}{611}\cdot\dfrac{121 + 51}{611} = 0.0792$

   b. $\dfrac{121}{121 + 279} = 0.303$

   f. $\dfrac{121 + 279}{611} + \dfrac{121 + 51}{611} - \dfrac{121}{611} = 0.738$

   c. $\dfrac{51}{51 + 160} = 0.242$

   d. $\dfrac{51}{51 + 121} = 0.297$

   g. $\dfrac{\left(\dfrac{121}{611}\right)}{\left(\dfrac{172}{611}\right)} = 0.703$

4. Because the vertical scale begins at 60 instead of 0, the difference between the two amounts is exaggerated. The graph makes it appear that men's earnings are roughly twice those of women, but men earn roughly 1.2 times the earnings of women.

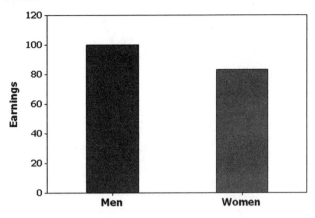

5. a. Frequency distribution or frequency table

    b. Probability distribution

    c. $\bar{x} = \dfrac{0 \cdot 9 + 1 \cdot 7 + 2 \cdot 12 + 3 \cdot 10 + 4 \cdot 10 + 5 \cdot 11 + 6 \cdot 8 + 7 \cdot 8 + 8 \cdot 14 + 9 \cdot 11}{9 + 7 + 12 + 10 + 10 + 11 + 8 + 8 + 14 + 11} = 4.7$

    This value is a statistic

    d. $\mu = 0 \cdot 0.1 + 1 \cdot 0.1 + 2 \cdot 0.1 + 3 \cdot 0.1 + 4 \cdot 0.1 + 5 \cdot 0.1 + 6 \cdot 0.1 + 7 \cdot 0.1 + 8 \cdot 0.1 + 9 \cdot 0.1 = 4.5$. This value is a parameter

    e. The random generation of 1000 digits should have a mean close to 4.5 from part (d). The mean of 4.5 is the mean for the population of all random digits; so samples will have means that tend to center about 4.5

6. a. $_{16}C_4 \cdot 0.1^4 \cdot 0.9^{12} = 0.0514$

    b. $1 - {}_{16}C_0 \cdot 0.1^0 \cdot 0.9^{16} = 0.815$

    c. This is a voluntary response sample. This suggests that the results might not be valid, because those with a strong interest in the topic are more likely to respond.

Chapter 6: Normal Probability Distributions

Section 6-2

1. The word "normal" has a special meaning in statistics. It refers to a specific bell-shaped distribution that can be described by Formula 6-1.

3. The mean and standard deviation have values of $\mu = 0$ and $\sigma = 1$

5. $0.2(5-1.25) = 0.75$

9. $P(z < 0.44) = 0.6700$

7. $0.2(3-1) = 0.40$

11. $P(-0.84 < z < 1.28) = P(z < 1.28) - P(z < -0.84) = 0.8997 - 0.2005 = 0.6992$ (Tech: 0.6993)

13. $z = 1.23$

21. $P(z > 0.82) = 1 - 0.7939 = 0.2061$

15. $z = -1.45$

23. $P(z > -1.50) = 1 - 0.0668 = 0.9332$

17. $P(z < -2.04) = 0.0207$

19. $P(z < 2.33) = 0.9901$

25. $P(0.25 < z < 1.25) = P(z < 1.25) - P(z < 0.25) = 0.8944 - 0.5987 = 0.2957$ (Tech: 0.2956)

27. $P(-2.75 < z < -2.00) = P(z < -2.00) - P(z < -2.75) = 0.0228 - 0.0030 = 0.0198$

29. $P(-2.20 < z < 2.50) = P(z < 2.50) - P(z < -2.20) = 0.9938 - 0.0139 = 0.9799$

31. $P(-2.11 < z < 4.00) = P(z < 4.00) - P(z < -2.11) = 0.9999 - 0.0174 = 0.9825$ (Tech: 0.9827)

33. $P(z < 3.65) = 0.9999$

39. $P_{2.5} = -1.96$ and $P_{97.5} = 1.96$

35. $P(z < 0) = 0.5000$

41. $z_{0.025} = 1.96$

37. $P_{90} = 1.28$

43. $z_{0.05} = 1.645$

45. $P(-1 < z < 1) = P(z < 1) - P(z < -1) = 0.8413 - 0.1587 = 0.6826 = 68.26\%$ (Tech: 68.27%)

47. $P(-3 < z < 3) = P(z < 3) - P(z < -3) = 0.9987 - 0.0013 = 0.9974 = 99.74\%$ (Tech: 99.73%)

49. a. $P(-1 < z < 1) = P(z < 1) - P(z < -1) = 0.8413 - 0.1587 = 0.6826 = 68.26\%$ (Tech: 68.27%)

b. $P(z < -2 \text{ or } z > 2) = P(z < -2) + P(z > 2) = 0.0228 + 0.0228 = 0.0456 = 4.56\%$

c. $P(-1.96 < z < 1.96) = P(z < 1.96) - P(z < -1.96) = 0.975 - 0.020 = 0.9500 = 95\%$

d. $P(-2 < z < 2) = P(z < 2) - P(z < -2) = 0.9772 - 0.0228 = 0.9544 = 95.44\%$ (Tech: 95.45%)

e. $P(z > 3) = 1 - P(z < 3) = 1 - 0.9987 = 0.0013 = 0.13\%$

Section 6-3

1. a. $\mu = 0$ and $\sigma = 1$

b.   The z scores are numbers without units of measurements

3. The standard normal distribution has a mean of 0 and a standard deviation of 1, but a nonstandard normal distribution has a different value for one or both of those parameters.

5. $z_{x=118} = \dfrac{118-100}{15} = 1.2$ which has an area of 0.8849 to the left of it

7. $z_{x=133} = \dfrac{133-100}{15} = 2.2$ which has an area of 0.9861 to the left of it. $z_{x=79} = \dfrac{110-100}{15} = -1.4$ which has

an area of 0.0808 to the left of it. The area between the two scores is $0.9861 - 0.0808 = 0.9053$.

9. $z = 2.44$, which means $x = 2.44 \cdot 15 + 100 = 136$

11. $z = -2.07$, which means $x = -2.07 \cdot 15 + 100 = 69$

13. $z_{x=85} = \dfrac{85-100}{15} = -1$, which has an area of 0.1587 to the left of it

15. $z_{x=90} = \dfrac{90-100}{15} = -0.67$ which has an area of 0.2514 to the left of it. $z_{x=110} = \dfrac{110-100}{15} = 0.67$ which

has an area of 0.7486 to the left of it. The area between the two scores is $0.7486 - 0.2514 = 0.4972$.
(Tech: 0.4950)

17. $z = 1.27$ which means the score is $x = 1.27 \cdot 15 + 100 = 119$

19. $z = 0.67$ which means the score is $x = 0.67 \cdot 15 + 100 = 110$

21. a. $z_{x=78} = \dfrac{78-63.8}{2.6} = 5.46$ which has an area of 0.9999 to the left of it.

$z_{x=62} = \dfrac{62-63.8}{2.6} = -0.69$ which has an area of 0.2451 to the left of it. Therefore, the percentage of

qualified women is $0.9999 - 0.2451 = 0.7548$ or 75.48%. (Tech 95.56%.) Yes, about 25% of women
are not qualified because of their heights.

b. $z_{x=78} = \dfrac{78-69.5}{2.4} = 3.54$ which has an area of 0.9999 to the left of it. $z_{x=62} = \dfrac{62-69.5}{2.4} = -3.13$

which has an area of 0.0009 to the left of it. Therefore, the percentage of men is
$0.9999 - 0.0009 = 0.9990$ or 99.90%. (Tech: 99.89%.) No, only about 0.1% of men are not qualified
because of their heights.

c. The z score with 2% to the left of it is –2.04 which corresponds to a height
of $x = -2.04 \cdot 2.6 + 63.8 = 58.5$ in. The z score with 2% to the right of it is 2.04 which corresponds to
a height of $x = 2.04 \cdot 2.6 + 63.8 = 69.1$ in.

d. The z score with 1% to the left of it is –2.33 which corresponds to a height of
$x = -2.33 \cdot 2.4 + 69.5 = 63.9$ in. The z score with 1% to the right of it is 2.33 which corresponds to a
height of $x = 2.33 \cdot 2.4 + 69.5 = 75.1$ in.

23. a. The height minimum is $4 \cdot 12 + 8 = 56$ in. and the height maximum is $6 \cdot 12 + 3 = 75$ in. The z score for

women for the minimum is $\dfrac{56-63.8}{2.6} = -3$, and the z score for women for the maximum is

$\dfrac{75-63.8}{2.6} = 4.31$. The area between the z scores is $0.9999 - 0.0013 = 0.9986$ or 99.86%

b. The z score for men for the minimum is $\dfrac{56-69.5}{2.4} = -5.63$, and the z score for men for the maximum

is $\dfrac{56-69.5}{2.4} = 2.29$. The area between the z scores is $0.9890 - 0.0001 = 0.9898$ or 98.98%. (Tech:
98.90%.)

23. (continued)

c. The z score with 5% for women to the left of it is –1.65 which corresponds to a height of $-1.65 \cdot 2.6 + 63.8 = 59.5$ in. the z score with 5% of men to the right of it is 1.625 which corresponds to a height of $1.625 \cdot 2.4 + 69.5 = 73.4$ in.

25. a. The z score for 174 lb. is $\dfrac{174 - 182.9}{40.8} = -0.22$ which has an area of 0.4129 to the left of it. (Tech: 0.4137.)

b. $\dfrac{3500}{140} = 25$ people

c. $\dfrac{3500}{182.9} = 19.14$, so 19 people

d. The mean weight is increasing over time, so safety limits must be periodically updated to avoid an unsafe condition

27. a. The z score for a 308 day pregnancy is $\dfrac{308 - 268}{15} = 2.67$ which corresponds to a probability of 0.0038 or 0.38%. Either a very rare event occurred or the husband is not the father.

b. The z score corresponding to 3% is –1.87 which corresponds to a pregnancy of $-1.87 \cdot 15 + 268 = 240$ days

29. a. The z score for an earthquake of magnitude 2 is $\dfrac{2 - 1.184}{0.587} = 1.39$ which is 0.9177 or 91.77% of earthquakes. (Tech: 99.78%.)

b. The z score is $\dfrac{4 - 1.184}{0.587} = 4.80$ which is 0.0001 or 0.01% of earthquakes. (Tech: 0.00%.)

c. The z score for 95% of earthquakes is 1.645 or 1.65 which corresponds to an earthquake magnitude of $1.645 \cdot 0.587 + 1.184 = 2.15$, so not all earthquakes about the 95th percentile will cause items to shake.

31. The z score for P_1 is –2.33 which corresponds to a count of $-2.33 \cdot 2.6 + 24 = 17.9$ chocolate chips. (Tech: 18 chocolate chips.) The z score for P_{99} is 2.33 which corresponds to a count of $2.33 \cdot 2.6 + 24 = 30.1$ chocolate chips. (Tech: 10.0 chocolate chips.) The values can be used to identify cookies with an unusually low number of chocolate chips or an unusually high number of chocolate chips, so those numbers can be used to monitor the production process to ensure that the numbers of chocolate chips stay within reasonable limits.

33. a. The mean is 67.25 beats per minute and the standard deviation is 10.335 beats per minute. The histogram for the data confirms that the distribution is roughly normal.

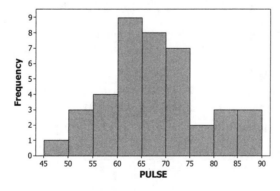

b. The z score for the bottom 2.5% is –1.95 which corresponds to a pulse of $-1.95 \cdot 10.335 + 67.25 = 47$ beats per minute, and the z score for the top 2.5% is 1.95 which corresponds to a pulse of $1.95 \cdot 10.335 + 67.25 = 87.5$ beats per minute.

35. a.   The new mean is equal to the old one plus the new points which is 75. The standard deviation is unchanged at 10 (since we added the same amount to each student.)

    b.   No, the conversion should also account for variation.

    c.   The z score for the bottom 70% is 0.52 which has a corresponding score of $0.52 \cdot 10 + 40 = 45.2$, and the z score for the top 10% is 1.28 which has a corresponding score of $1.28 \cdot 10 + 40 = 52.8$

    d.   Using a scheme like the one in part (c), because variation is included in the curving process.

37. The z score for Q_1 is –0.67, and the z score for Q_3 is 0.67. The IQR is $0.67 - (-0.67) = 1.34$.

    $1.5 \cdot IQR = 2.01$, so $Q_1 - 1.5 \cdot IQR = -0.67 - 2.01 = -2.68$ and $Q_3 + 1.5 \cdot IQR = 0.67 + 2.01 = 2.68$

    The percentage to the left of –2.68 is 0.0037 and the percentage to the right of 2.68 is 0.0037. Therefore, the percentage of an outlier is 0.0074. (Tech: 0.0070)

Section 6-4

1. a.   The sample mean will tend to center about the population parameter of 5.67 g.

    b.   The sample mean will tend to have a distribution that is approximately normal.

    c.   The sample proportions will tend to have a distribution that is approximately normal.

3.   Sample mean, sample variance, sample proportion

5.   No. The sample is not a simple random sample from the population of all college Statistics students. It is very possible that the students at Broward College do not accurately reflect the behavior of all college Statistics students.

7. a.   The mean of the population is $\mu = \dfrac{4 + 5 + 9}{3} = 6$, and the variance is

    $$\sigma^2 = \frac{(4-6)^2 + (5-6)^2 + (9-6)^2}{3} = 4.7$$

    b.   The possible sample of size 2 are {(4, 4), (4, 5), (4, 9), (5, 4), (5, 5), (5, 9), (9, 4), (9, 5), (9, 9)} which have the following variances {0, 0.5, 12.5, 0.5, 0, 8, 12.5, 8, 0} respectively.

Sample Variance	Probability
0	3/9
0.5	2/9
8	2/9
12.5	2/9

    c.   The sample variances' mean is $\dfrac{3 \cdot 0 + 2 \cdot 0.5 + 2 \cdot 8 + 2 \cdot 12.5}{9} = 4.7$

    d.   Yes. The mean of the sampling distribution of the sample variances (4.7) is equal to the value of the population variance (4.7) so the sample variances target the value of the population variance.

9. a.   The population median is 5

    b.   The possible sample of size 2 are {(4, 4), (4, 5), (4, 9), (5, 4), (5, 5), (5, 9), (9, 4), (9, 5), (9, 9)} which have the following medians {4, 4.5, 6.5, 4.5, 5, 7, 6.5, 7, 9}

Sample Median	Probability
4	1/9
4.5	2/9
5	1/9
6.5	2/9
7	2/9
9	1/9

9. (continued)

   c. The mean of the sampling distribution of the sampling median is

$$\frac{4+4.5+4.5+5+6.5+6.5+7+7+9}{9}=6$$

   d. No. The mean of the sampling distribution of the sample medians is 6, and it is not equal to the value of the population median of 5, so the sample medians do not target the value of the population median.

11. a. The possible samples of size 2 are {(56, 56), (56, 49), (56, 58), (56, 46), (49, 56), (49, 49), (49, 58), (49, 46), (58, 56), (58, 49), (58, 58), (58, 46), (46, 56), (46, 49), (46, 58), (46, 46)}

Sample Mean Age	Probability
46	1/16
47.5	2/16
49	1/16
51	2/16
52	2/16
52.5	2/16
53.5	2/16
56	1/16
57	2/16
58	1/16

   b. The mean of the population is $\dfrac{56+49+58+46}{4}=52.25$ and the mean of the sample means is

$$\frac{46+47.5+47.5+49+51+51+52+52+52.5+52.5+53.5+53.5+56+57+57+58}{16}=52.25$$

   c. The sample means target the population mean. Sample means make good estimators of population means because they target the value of the population mean instead of systematically underestimating or overestimating it.

13. a. The possible samples of size 2 are {(56, 56), (56, 49), (56, 58), (56, 46), (49, 56), (49, 49), (49, 58), (49, 46), (58, 56), (58, 49), (58, 58), (58, 46), (46, 56), (46, 49), (46, 58), (46, 46)} which have the following ranges and associated probabilities

Sample Range	Probability
0	4/16
2	2/16
3	2/16
7	2/16
9	2/16
10	2/16
12	2/16

   b. The range of the population is $58-46=12$, the mean of the sample ranges is

$$\frac{4\cdot0+2\cdot2+2\cdot3+2\cdot7+2\cdot9+2\cdot10+2\cdot12}{16}=5.375.$$ The values are not equal.

   c. The sample ranges do not target the population range of 12, so sample ranges do not make good estimators of the population range.

15. The possible birth samples are {(b, b), (b, g), (g, b), (g, g)}

Proportion of Girls	Probability
0	0.25
1 / 2	0.5
2/2	0.25

Yes. The proportion of girls in 2 births is 0.5, and the mean of the sample proportions is 0.5. The result suggests that a sample proportion is an unbiased estimator of the population proportion.

17. The possibilities are: both questions incorrect, one question correct (two choices), both questions correct.

a.

Proportion Correct	Probability
$\dfrac{0}{2}$	$\dfrac{4}{5} \cdot \dfrac{4}{5} = \dfrac{16}{25}$
$\dfrac{1}{2}$	$2 \cdot \left(\dfrac{1}{5} \cdot \dfrac{4}{5} \right) = \dfrac{8}{25}$
$\dfrac{2}{2}$	$\left(\dfrac{1}{5} \cdot \dfrac{1}{5} \right) = \dfrac{1}{25}$

b.   The mean is $\dfrac{16 \cdot 0 + 8 \cdot 0.5 + 1 \cdot 1}{25} = 0.2$

c.   Yes. The sampling distribution of the sample proportions has a mean of 0.2 and the population proportion is also 0.2 (because there is 1 correct answer among 5 choices.) Yes, the mean of the sampling distribution of the sample proportions is always equal to the population proportion.

19. $P(0) = \dfrac{1}{2(2 - 2 \cdot 0)!(2 \cdot 0)!} = \dfrac{1}{4} = 0.25$, $P(0.5) = \dfrac{1}{2(2 - 2 \cdot 0.5)!(2 \cdot 0.5)!} = 0.5$,

$P(1) = \dfrac{1}{2(2 - 2 \cdot 1)!(2 \cdot 1)!} = 0.25$. The formula yields values which describes the sampling distribution of the sample proportions. The formula is just a different way of presenting the same information in the table that describes the sampling distribution.

Section 6-5

1.   Because the sample size is greater than 30, the sampling distribution of the mean ages can be approximated by a normal distribution with mean μ and standard deviation $\dfrac{\sigma}{\sqrt{40}}$.

3. $\mu_{\bar{x}} = 60.5 \, \text{cm}$ and it represents the mean of the population consisting of all sample means.

$\sigma_{\bar{x}} = \dfrac{6.6}{\sqrt{36}} = 1.1 \, \text{cm}$, and it represents the standard deviation of the population consisting of all sample means.

5.   a. $z_{x = 222.7} = \dfrac{222.7 - 205.5}{8.6} = 2$, which has a probability of 0.9772.

b. $z_{x = 207} = \dfrac{207 - 205.5}{\dfrac{8.6}{\sqrt{49}}} = 1.22$, which has a probability of 0.8888. (Tech: 0.8889.)

7. a. $z_{x=218.4} = \dfrac{218.4 - 205.5}{8.6} = 1.5$, which has a probability of $1 - 0.9332 = 0.0668$ to the right of it

   b. $z_{x=204} = \dfrac{204 - 205.5}{\frac{8.6}{\sqrt{9}}} = -0.52$ which has a probability of $1 - 0.3015 = 0.6985$ to the right of it. (Tech: 0.6996.)

   c. Because the original population has a normal distribution, the distribution of sample means is normal for any sample size.

9. a. $z_{x=231.5} = \dfrac{231.5 - 205.5}{8.6} = 3.02$ and $z_{x=179.7} = \dfrac{179.7 - 205.5}{8.6} = -3$ which has a probability of $0.9987 - 0.0013 = 0.9974$ between them. (Tech: 0.9973.)

   b. $z_{x=206} = \dfrac{206 - 205.5}{\frac{8.6}{\sqrt{40}}} = 0.37$ and $z_{x=204} = \dfrac{204 - 205.5}{\frac{8.6}{\sqrt{40}}} = -1.1$ which has a probability of $0.6443 - 0.1357 = 0.5086$ between them. (Tech: 0.5085.)

11. $z_{x=195.3} = \dfrac{195.3 - 182.9}{\frac{40.8}{\sqrt{16}}} = 1.22$ which has a probability of $1 - 0.8888 = 0.1112$ to the right of it. The elevator appears to be relatively safe because there is a very small chance that it will be overloaded with 16 male passengers. (Tech: 0.1121.)

13. a. $z_{x=25} = \dfrac{25 - 22.65}{0.8} = 2.94$ and $z_{x=21} = \dfrac{21 - 22.65}{0.8} = -2.06$, so $0.9984 - 0.0197 = 0.9787 = 97.87\%$ of women can fit into the hats. (Tech: 0.9788.)

   b. The z scores for the smallest 2.5% and the largest 2.5% head circumferences are –1.96 and 1.96 respectively. This corresponds to head circumferences of $0.8 \cdot (-1.96) + 22.65 = 21.08$ and $0.8 \cdot 1.96 + 22.65 = 24.22$

   c. $z_{x=23} = \dfrac{23 - 22.65}{\frac{0.8}{\sqrt{64}}} = 3.5$ and $z_{x=22} = \dfrac{22 - 22.65}{\frac{0.8}{\sqrt{64}}} = -6.5$ which have a probability of $0.9998 - 0.0000 = 0.9998 = 99.98\%$ between them. No, the hats must fit individual women, not the mean from 64 women. If all hats are made to fit head circumferences between 22 in. and 23 in., the hats will not fit about half those women.

15. a. The mean weight of passengers is $\dfrac{3500}{25} = 140$ lb.

   b. $z_{x=140} = \dfrac{140 - 182.9}{\frac{40.8}{\sqrt{25}}} = -5.26$ which has a probability of 0.99999 (or 1.0000) to the right of it. (Tech: 0.0.99999993.)

   c. $z_{x=175} = \dfrac{175 - 182.9}{\frac{40.8}{\sqrt{20}}} = -0.87$ which has a probability of 0.8078 to the right of it. (Tech: 0.8067.)

   d. Given that there is a 0.8078 probability of exceeding the 3500 lb. limit when the water taxi is loaded with 20 random men, the new capacity of 20 passengers does not appear to be safe enough because the probability of overloading is too high.

17. a. $z_{x=167} = \dfrac{167-182.9}{40.8} = -0.39$ which has a probability of $1-0.3483 = 0.6517$ to the right of it.

    (Tech: 0.6516.)

    b. $z_{x=167} = \dfrac{167-182.9}{\dfrac{40.8}{\sqrt{12}}} = -1.35$ which has a probability of $1-0.0885 = 0.9115$ to the right of it

    c. There is a high probability that the gondola will be overloaded if it is occupied by 12 more people, so it appears that the number of allowed passengers should be reduced.

19. a. $z_{x=211} = \dfrac{211-165}{45.6} = 1.01$ and $z_{x=140} = \dfrac{140-165}{45.6} = -0.55$ which have a probability of

    $0.8438 - 0.2912 = 0.5526$ between them. (Tech: 0.5517.)

    b. $z_{x=211} = \dfrac{211-165}{\dfrac{45.6}{\sqrt{36}}} = 6.05$ and $z_{x=140} = \dfrac{140-165}{\dfrac{45.6}{\sqrt{36}}} = -3.29$ which have a probability of

    $0.9999 - 0.0005 = 0.9994$ between them. (Tech: 0.9995.)

    c. Part (a) because the ejection seats will be occupied by individual women, not groups of women.

21. a. $z_{x=72} = \dfrac{72-69.5}{2.4} = 1.04$ which has a probability of 0.8508. (Tech: 0.8512.)

    b. $z_{x=72} = \dfrac{72-69.5}{\dfrac{2.4}{\sqrt{100}}} = 14.76$ which has a probability of 0.9999.

    (Tech: 1.0000 when rounded to four decimal places.)

    c. The probability of Part (a) is more relevant because it shows that 85.08% of male passengers will not need to bend. The result from part (b) gives us useful information about the comfort and safety of individual male passengers.

    d. Because men are generally taller than women, a design that accommodates a suitable proportion of men will necessarily accommodate a greater proportion of women.

23. a. Yes. The sampling is without replacement and the sample size of 50 is greater than 5% of the finite

    population size of 275. $\sigma_{\bar{x}} = \dfrac{16}{\sqrt{50}}\sqrt{\dfrac{275-50}{275-1}} = 2.0504584$

    b. $z_{x=105} = \dfrac{105-95.5}{2.0504584} = 4.63$ and $z_{x=95.5} = \dfrac{95-95.5}{2.0504584} = -0.24$ which have a probability of

    $1-0.4053 = 0.5947$. (Tech: 0.5963.)

25. a. $\mu = \dfrac{4+5+9}{3} = 6$, $\sigma = \sqrt{\dfrac{(4-6)^2 + (5-6)^2 + (9-6)^2}{3}} = 2.160246899$

    b. The possible samples of size 2 are:$\{\,(4, 5),\ (4, 9),\ (5, 4),\ (5, 9),\ (9, 4),\ (9, 5)\}$ which have the following means $\{4.5,\ 6.5, 4.5, 7, 6.5, 7\}$ respectively.

    c. $\mu_{\bar{x}} = \dfrac{4.5+6.5+4.5+7+6.5+7}{6} = 6$ and

    $\sigma_{\bar{x}} = \sqrt{\dfrac{(4.5-6)^2 + (6.5-6)^2 + (4.5-6)^2 + (7-6)^2 + (6.5-6)^2 + (7-6)^2}{6}} = 1.08012345$

    d. It is clear that $\mu = \mu_{\bar{x}} = 6$. $\sigma_{\bar{x}} = \dfrac{2.160246899}{\sqrt{2}}\sqrt{\dfrac{3-2}{3-1}} = 1.08012345 = \sigma$

Section 6-6

1.   The histogram should be approximately bell-shaped, and the normal quantile plot should have points that approximate a straight line pattern.

3.   We must verify that the sample is from a population having a normal distribution. We can check for normality using a histogram, identifying the number of outliers, and constructing a normal quantile plot.

5.   Not normal. The points show a systematic pattern that is not a straight line pattern.

7.   Normal. The points are reasonably close to a straight line pattern, and there is no other pattern that is not a straight line pattern.

9.   Not normal

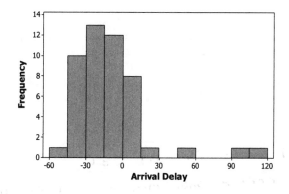

11. Normal

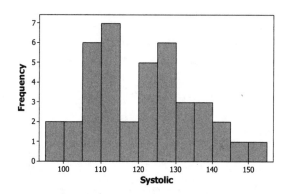

13. Not normal

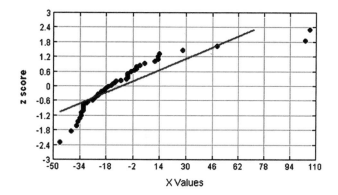

15. Normal

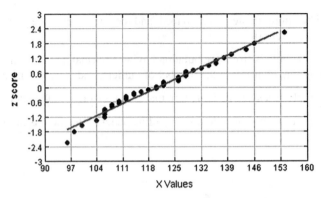

17. Normal. The points have coordinates (–131, –1.28), (134, –0.52), (139, 0), (143, 0.52), (145, 1.28)

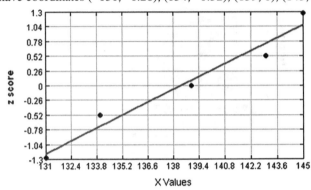

19. Not normal. The points have coordinates (1034, –1.53), (1051, –0.89), (1067, –0.49), (1070, –0.16), (1079, 0.16), (1079, 0.49), (1173, 0.89), (1272, 1.53)

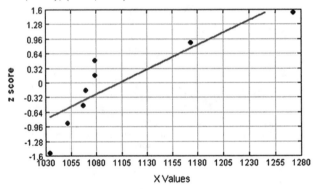

21. a.   Yes.
     b.   Yes.
     c.   No.

23. The original values are not from a normally distributed population.

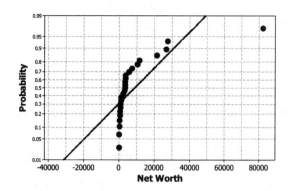

After taking the logarithm of each value, the values appear to be from a normally distributed population.

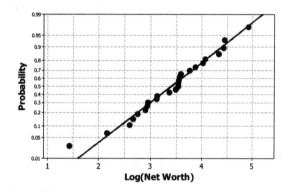

The original values are from a population with a lognormal distribution.

Section 6-7

1. The Minitab display shows that the region representing 235 wins is a rectangle. The result of 0.0068 is an approximation, but the result of 0.0066 is better because it is based on an exact calculation. The approximation differs from the exact result by a very small amount.

3. $p = \dfrac{1}{5} = 0.2$, $q = \dfrac{4}{5} = 0.8$, $\mu = 25 \cdot 0.2 = 5$, $\sigma = \sqrt{25 \cdot 0.2 \cdot 0.8} = 2$. The value of 5 for the mean shows that for many people who make random guesses for the 25 questions, the mean number of correct answers is 5. For many people who make random guesses, the standard deviation of 2 is a measure of how much the numbers of correct responses vary.

5. The requirements are satisfied with a mean of $13 \cdot 0.4 = 5.2$ and the standard deviation of $\sqrt{13 \cdot 0.4 \cdot 0.6} = 1.766$. Therefore, $z_{x=2.5} = \dfrac{2.5 - 5.2}{\sqrt{13 \cdot 0.4 \cdot 0.6}} = -1.53$ which has a probability of 0.0630. (Tech: 0.0632.)

7. The requirement of $nq \ge 5$ is not satisfied. Normal approximation should not be used.

9. $\mu = 100 \cdot 0.22 = 22$, $\sigma = \sqrt{100 \cdot 0.22 \cdot 0.78} = 4.1425$

$z_{x=19.5} = \dfrac{19.5 - 22}{\sqrt{100 \cdot 0.22 \cdot 0.78}} = -0.60$ which has a probability of 0.2743. (Tech: 0.2731.)

11. $z_{x=22.5} = \dfrac{22.5 - 22}{\sqrt{100 \cdot 0.22 \cdot 0.78}} = 0.12$ and $z_{x=23.5} = \dfrac{23.5 - 22}{\sqrt{100 \cdot 0.22 \cdot 0.78}} = 0.36$ which have a probability of

$0.6406 - 0.5478 = 0.0928$ between them. (Tech: 0.0933.)

13. $\mu = 611 \cdot 0.3 = 183.3$, $\sigma = \sqrt{611 \cdot 0.3 \cdot 0.7} = 11.3274$

    a. $z_{x=172.5} = \dfrac{172.5 - 183.3}{\sqrt{611 \cdot 0.3 \cdot 0.7}} = -0.95$ and $z_{x=171.5} = \dfrac{171.5 - 183.3}{\sqrt{611 \cdot 0.3 \cdot 0.7}} = -1.04$ which have a probability of

         $0.1711 - 0.1492 = 0.0219$ between them. (Tech using normal approximation: 0.0214; Tech using binomial: 0.0217)

    b. $z_{x=172.5} = \dfrac{172.5 - 183.3}{11.3274} = -0.95$ which has a probability of 0.1711. The result of 172 overturned calls

         is not unusually low. (Tech using normal approximation: 0.1702; Tech using binomial: 0.1703.)

    c.   The result from part (b) is useful. We want the probability of getting a result that is at least as extreme as the one obtained.

    d.   If the 30% rate is correct, there is a good chance (17.11%) of getting 172 or fewer calls overturned, so there is not strong evidence against the 30% rate.

15. $\mu = 580 \cdot 0.75 = 435$, $\sigma = \sqrt{580 \cdot 0.75 \cdot 0.25} = 10.4283$

    a. $z_{x=428.5} = \dfrac{428.5 - 435}{\sqrt{580 \cdot 0.75 \cdot 0.25}} = -0.62$ and $z_{x=427.5} = \dfrac{427.5 - 435}{\sqrt{580 \cdot 0.75 \cdot 0.25}} = -0.72$ which have a probability

         of $0.2676 - 0.2358 = 0.0318$ between them . (Tech using normal approximation: 0.0305; Tech using binomial: 0.0301.)

    b. $z_{x=428.5} = \dfrac{428.5 - 435}{\sqrt{580 \cdot 0.75 \cdot 0.25}} = -0.62$ which has a probability of 0.2676. The result of 428 peas with

         green pods is not unusually low. (Tech using normal approximation: 0.2665; Tech using binomial: 0.2650.)

    c.   The result from part (b) is useful. We want the probability of getting a result that is at least as extreme as the one obtained.

    d.   No. Assuming that Mendel's probability of 3/4 is correct, there is a good chance (26.76%) of getting the results that were obtained. The obtained results do not provide strong evidence against the claim that the probability of a pea having a green pod is 3/4

17. $\mu = 945 \cdot 0.5 = 472.5$, $\sigma = \sqrt{945 \cdot 0.5 \cdot 0.5} = 15.3704$

    a. $z_{x=879.5} = \dfrac{879.5 - 472.5}{\sqrt{945 \cdot 0.5 \cdot 0.5}} = 26.48$ and $z_{x=878.5} = \dfrac{878.5 - 472.5}{\sqrt{945 \cdot 0.5 \cdot 0.5}} = 26.41$ which have a probability of

         0.0000 or 0+ (a very small positive probability that is extremely close to 0) between them.

    b. $z_{x=878.5} = \dfrac{878.5 - 472.5}{\sqrt{945 \cdot 0.5 \cdot 0.5}} = 26.41$ which has a probability of 0.0001. (Tech: 0.0000 or 0+, which is a

         very small positive probability that is extremely close to 0). If boys and girls are equally likely, 879 girls in 945 births is unusually high.

    c.   The result from part (b) is more relevant, because we want the probability of a result that is at least as extreme as the one obtained.

    d.   Yes. It is very highly unlikely that we would get 879 girls in 945 births by chance. Given that the 945 couples were treated with the XSORT method, it appears that this method is effective in increasing the likelihood that a baby will be a girl.

19. $\mu = 1002 \cdot 0.61 = 611.22$, $\sigma = \sqrt{1002 \cdot 0.61 \cdot 0.39} = 15.4394$

$z_{x=700.5} = \dfrac{700.5 - 611.22}{\sqrt{1002 \cdot 0.61 \cdot 0.39}} = 5.78$ which has a probability of 0.0001 to the right of it. (Tech 0.0000.) The result suggests that the surveyed people did not respond accurately.

21. The probability of six or fewer should be computed. $\mu = 50 \cdot 0.2 = 10$, $\sigma = \sqrt{50 \cdot 0.2 \cdot 0.8} = 2.8284$

$z_{x=6.5} = \dfrac{6.5 - 10}{\sqrt{50 \cdot 0.2 \cdot 0.8}} = -1.24$ which has a probability of 0.1075. (Tech using normal approximation: 0.1080; Tech using binomial: 0.1034.) Because that probability is not very small, the evidence against the rate of 20% is not very strong.

23. The probability of 170 or fewer should be computed. $\mu = 1000 \cdot 0.20 = 200$, $\sigma = \sqrt{1000 \cdot 0.2 \cdot 0.8} = 12.6491$

$z_{x=170.5} = \dfrac{170.5 - 200}{\sqrt{1000 \cdot 0.2 \cdot 0.8}} = -2.33$ which has a probability of 0.0099. (Tech using normal approximation: 0.0098; Tech using binomial: 0.0089.) Because the probability of 170 or fewer is so small with the assumed 20% rate, it appears that the rate is actually less than 20%.

25. a. In order to make a profit Marc will need to win over $1000. With 35:1 odds a $5 bet wins $175. Therefore, Marc needs 6 winning bets in order to make a profit.

$\mu = 200 \cdot \dfrac{1}{38} = 5.2632$, $\sigma = \sqrt{200 \cdot \dfrac{1}{38} \cdot \dfrac{37}{38}} = 2.2638$

$z_{x=5.5} = \dfrac{5.5 - 5.2632}{2.2638} = 0.10$ which has a probability of $1.0000 - 0.5398 = 0.4602$ to the right of it.

(Tech using normal approximation: 0.4583; tech using binomial: 0.4307)

   b. Since the odds of winning are 1:1 Marc would need 101 wins or more to make a

profit. $\mu = 200 \cdot \dfrac{244}{495} = 98.5859$, $\sigma = \sqrt{200 \cdot \dfrac{244}{495} \cdot \dfrac{251}{495}} = 7.0704$

$z_{x=100.5} = \dfrac{100.5 - 98.5859}{7.0704} = 0.27$ which has a probability of $1.0000 - 0.6064 = 0.3936$

(Tech using normal approximation: 0.3933; tech using binomial: 0.3932)

   c. The roulette game provides a better likelihood of making a profit.

Chapter Quick Quiz

1. $\mu = 0$ and $\sigma = 1$

2.

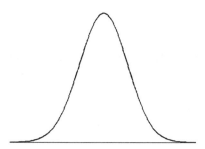

3. $P_{98} = 2.05$ (Tech: 2.05375)

4. $P(z > -1) = 1 - P(z < -1) = 1 - 0.1587 = 0.8413$

5. $P(1.37 < z < 2.42) = P(z < 2.42) - P(z < 1.37) = 0.9922 - 0.9147 = 0.0775$ (Tech: 0.0076)

6. $z_{x=4.2} = \dfrac{4.2 - 4.577}{0.382} = -0.99$ which have a probability of 0.1611. (Tech: 0.1618.)

7. $z_{x=5.4} = \dfrac{5.4 - 4.577}{0.382} = 2.15$ which has a probability of $1.0000 - 0.9842 = 0.0158$. (Tech: 0.0156.)

8.   The z score for $P_{80} = 0.84$ which corresponds to a red blood count of $0.84 \cdot 0.382 + 4.577 = 4.898$

9. $z = \dfrac{4.444 - 4.577}{\dfrac{0.382}{\sqrt{25}}} = -1.74$ which has a probability of 0.0409

10. $z_{x=4.2} = \dfrac{4.2 - 4.577}{0.382} = -0.99$ and $z_{x=5.4} = \dfrac{5.4 - 4.577}{0.382} = 2.15$ which have a probability of $0.9842 - 0.1611 = 0.8231$ or 82.31%. (Tech: 82.26%)

Review Exercises

1.   a.   The probability to the left of a z score of 2.93 is 0.9983

     b.   The probability to the right of a z score of −1.53 is $1.0000 - 0.0630 = 0.9370$

     c.   The probability between z scores −1.07 and 2.07 is $0.9808 - 0.1423 = 0.8385$

     d.   The z score for $P_{30} = -0.52$

     e. $z_{x=0.27} = \dfrac{0.27 - 0}{\dfrac{1}{\sqrt{16}}} = 1.08$ which has a probability of $1 - 0.8599 = 0.1401$

2.   a. $z_{x=1605} = \dfrac{1605 - 1516}{63} = 1.41$ which has a probability of $1 - 0.9207 = 0.0793$ or 7.93%. (Tech: 7.89%.)

     b.   The z score for the lowest 1% is −2.33 which corresponds to a standing eye height of $-2.33 \cdot 63 + 1516 = 1369.2$ mm. (Tech: 1369.4.)

3.   a. $z_{x=1500} = \dfrac{1500 - 1634}{66} = -2.03$ which has a probability of $1 - 0.0212 = 0.9788$ or 97.88%

     b.   The z score for the lowest 95% is 1.645 which corresponds to a standing eye height of $1.645 \cdot 66 + 1634 = 1742.6$ mm

4.   a.   Normal distribution

     b. $\mu_{\bar{x}} = 21.1$

     c. $\sigma_{\bar{x}} = \dfrac{5.1}{\sqrt{80}} = 0.57$

5.   a.   An unbiased estimator is a statistic that targets the value of the population parameter in the sense that the sampling distribution of the statistic has a mean that is equal to the mean of the corresponding parameter.

     b.   Mean, variance and proportion

     c.   True

6.   a. $z_{x=72} = \dfrac{72 - 69.5}{2.4} = 1.04$ which has a probability of 0.8508 or 85.08%. (Tech: 85.12.) With about 15% of all men needing to bend, the design does not appear to be adequate, but the Mark VI monorail appears to be working quite well in practice.

     b.   The z score for 99% is 2.33 which corresponds to a doorway height of $2.33 \cdot 2.4 + 69.5 = 75.1$

7. a. $z_{x=175} = \dfrac{175 - 182.9}{40.9} = -0.19$ which has a probability of $1 - 0.4247 = 0.5753$. (Tech: 0.5766.)

   b. $z_{x=175} = \dfrac{175 - 182.9}{\dfrac{40.9}{\sqrt{213}}} = -2.82$ which has a probability of $1 - 0.0024 = 0.9976$. Yes, if the plane is full

   of male passengers, it is highly likely that it is overweight.

8. a. No. A histogram is far from bell shaped.

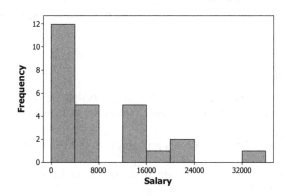

   b. No. The sample has a size of 26 which does not satisfy the condition at least 30, and the values do not appear to be from a population having a normal distribution.

9. $\mu = 1064 \cdot \dfrac{3}{4} = 798$, $\sigma = \sqrt{1064 \cdot \dfrac{3}{4} \cdot \dfrac{1}{4}} = 14.1244$

   $z_{z=787.5} = \dfrac{787.5 - 798}{\sqrt{1064 \cdot \dfrac{3}{4} \cdot \dfrac{1}{4}}} = -0.74$ which has a probability of 0.2296. (Tech using normal approximation:

   0.2286; Tech using binomial: 0.2278.) The occurrence of 787 offspring plants with long stem is not unusually low because its probability is not small. The results are consistent with Mendel's claimed proportion of 3/4

10. $\mu = 64 \cdot 0.8 = 51.2$, $\sigma = \sqrt{64 \cdot 0.8 \cdot 0.2} = 3.2$

    a. $z_{x=49.5} = \dfrac{49.5 - 51.2}{3.2} = -0.53$ which has a probability of $1 - 0.2981 = 0.7019$ to the right of it. (Tech using normal approximation: 0.7024; Tech using binomial: 0.7100)

    b. $z_{x=49.5} = -0.53$ and $z_{x=50.5} = \dfrac{50.5 - 51.2}{3.2} = -0.22$ which have a probability of

    $0.4129 - 0.2981 = 0.1148$ between them. (Tech using normal approximation: 0.1158; Tech using binomial: 0.1190)

Cumulative Review Exercises

1. a. $\bar{x} = \dfrac{14,500,000 + 145,000,000 + 14,000,000 + 5,000,000 + 3,500,000}{5} = \$10,300,000$

   b. The median is $14,000,000

   c. $s = \sqrt{\dfrac{(14,500 - 10,300)^2 + (14,500 - 10,300)^2 + ... + (5000 - 10,300)^2 + (3500 - 10,300)^2}{5 - 1}}$

   $= \$5552.027$ (in thousands of dollars) which is $5,552,027.

   d. $s^2 = 30,825,003,810,000$ square dollars

1.   (continued)

e. $z_{x=14,500,000} = \dfrac{14,500,000 - 10,300,000}{5,552,027} = 0.76$

f.   Ratio

g.   Discrete

h.   No, the starting players are likely to be the best players who receive the highest salaries.

2.   a. $\overline{A}$ is the event of selecting someone who does not have the belief that college is not a good investment. NOTE: This is not the same as selecting someone who believes that college is a good investment.

b. $P(\overline{A}) = 1 - 0.1 = 0.9$

c. $P = 0.1 \cdot 0.1 \cdot 0.1 = 0.001$

d.   The sample is a voluntary response sample. This suggests that the 10% rate might not be very accurate, because people with strong feelings or interest about the topic are more likely to respond.

3.   a. $z_{x=2500} = \dfrac{2500 - 3369}{567} = -1.53$ which has a probability of 0.0630. (Tech: 0.0627)

b.   The z score for the bottom 10% is −1.28, which correspond to the weight $-1.28 \cdot 567 + 3369 = 2642.24$ g. (Tech: 2642 g.)

c. $z_{x=1500} = \dfrac{1500 - 3369}{567} = -3.3$ which has a probability of 0.0005

d. $z_{x=3400} = \dfrac{3400 - 3369}{\dfrac{567}{\sqrt{25}}} = 0.27$ which has a probability of $1 - 0.6064 = 0.3936$. (Tech: 0.3923) to the

right of it.

4.   a.   The vertical scale does not start at 0, so differences are somewhat distorted. By using a scale ranging from 1 to 29 for frequencies that range 2 to 14, the graph is flattened, so differences are not shown as they should be.

b.   The graph depicts a distribution that is not exactly normal, but it is approximately normal because it is roughly bell shaped.

c.   Minimum: 42 years; maximum: 70 years. Using the range rule of thumb, the standard deviation is estimated to be $\dfrac{70 - 42}{4} = 7$ years. The estimate of 7 years is very close to the actual standard deviation of 6.6 years, so the range rule of thumb works quite well here.

5.   a. $P(X = 3) = 0.1 \cdot 0.1 \cdot 0.1 = 0.001$

b. $P(X \geq 1) = 1 - P(X = 0) = 1 - (0.9 \cdot 0.9 \cdot 0.9) = 0.271$

c.   The requirement that $np \geq 5$ is not satisfied, indicating that the normal approximation would result in errors that are too large.

d. $\mu = 50 \cdot 0.1 = 5$

e. $\sigma = \sqrt{50 \cdot 0.1 \cdot 0.9} = 2.1213$

f.   No, 8 is within two standard deviations of the mean and is within the range of values that could easily occur by chance.

Chapter 7: Estimates and Sample Sizes

Section 7-2

1. The confidence level (such as 95%) was not provided.

3. $\hat{p} = 0.26$ is the sample proportion; $\hat{q} = 0.74$ (found from evaluating $1 - \hat{p}$); $n = 1910$ is the sample size; $E = 0.03$ is the margin of error; p is the population proportion, which is unknown. The value of α is 0.05.

5. 1.28

7. 1.645

9. $E = \dfrac{0.186 - 0.0641}{2} = 0.061,$ so 0.125 ± 0.061

11. $0.0268 < p < 0.133$

13. a. $\hat{p} = \dfrac{531}{1002} = 0.530$

    b. $E = z_{\alpha/2}\sqrt{\dfrac{\hat{p}\hat{q}}{n}} = 1.96\sqrt{\dfrac{\left(\frac{531}{1002}\right)\left(\frac{471}{1002}\right)}{1002}} = 0.0309$

    c. $\hat{p} - E < p < \hat{p} - E \Rightarrow 0.530 - 0.0309 < p < 0.530 - 0.0309 \Rightarrow 0.499 < p < 0.561$

    d. We have 95% confidence that the interval from 0.499 to 0.561 actually does contain the true value of the population proportion.

15. a. $\hat{p} = \dfrac{1083}{2518} = 0.430$

    b. $E = z_{\alpha/2}\sqrt{\dfrac{\hat{p}\hat{q}}{n}} = 1.65\sqrt{\dfrac{\left(\frac{1083}{2518}\right)\left(\frac{1435}{2518}\right)}{2518}} = 0.0162$

    $$\hat{p} - E < p < \hat{p} - E$$

    c. $0.430 - 0.0162 < p < 0.430 - 0.0162$

    $$0.414 < p < 0.446$$

    d. We have 90% confidence that the interval from 0.414 to 0.446 actually does contain the true value of the population proportion.

17. a. $\hat{p} = \dfrac{879}{945} = 0.930$

    b. $\hat{p} \pm z_{\alpha/2}\sqrt{\dfrac{\hat{p}\hat{q}}{n}} = \dfrac{879}{945} \pm 1.96\sqrt{\dfrac{\left(\frac{879}{945}\right)\left(\frac{66}{945}\right)}{945}}$

    $$0.914 < p < 0.946$$

    c. Yes. The true proportion of girls with the XSORT method is substantially greater than the proportion of (about) 0.5 that is expected when no method of gender selection is used.

19. a. 0.5

    b. $\hat{p} = \dfrac{123}{280} = 0.439$

    c. $\hat{p} \pm z_{\alpha/2}\sqrt{\dfrac{\hat{p}\hat{q}}{n}} = \dfrac{123}{280} \pm 2.56\sqrt{\dfrac{\left(\frac{123}{280}\right)\left(\frac{157}{280}\right)}{280}}$

    $0.363 < p < 0.515$ or $36.3\% < p < 51.5\%$

19. (continued)

    d.   If the touch therapists really had an ability to select the correct hand by sensing an energy field, their success rate would be significantly greater than 0.5, but the sample success rate of 0.439 and the confidence interval suggest that they do not have the ability to select the correct hand by sensing an energy field.

21. a. $427(0.29) = 124$

    b. $\hat{p} \pm z_{\alpha/2}\sqrt{\dfrac{\hat{p}\hat{q}}{n}} = 0.29 \pm z_{0.025}\sqrt{\dfrac{(0.29)(0.71)}{427}}$

       $0.247 < p < 0.333$ or $24.7\% < p < 33.3\%$

    c.   Yes. Because all values of the confidence interval are less than 0.5, the confidence interval shows that the percentage of women who purchase books online is very likely less than 50%.

    d.   No. The confidence interval shows that it is possible that the percentage of women who purchase books online could be less than 25%.

    e.   Nothing.

23. a. $514(0.459) = 236$

    b. $\hat{p} \pm z_{\alpha/2}\sqrt{\dfrac{\hat{p}\hat{q}}{n}} = 0.459 \pm z_{0.025}\sqrt{\dfrac{(0.459)(0.541)}{514}}$ (using $x = 236$: $0.403 < p < 0.516$).

          $0.402 < p < 0.516$

    c. $\hat{p} \pm z_{\alpha/2}\sqrt{\dfrac{\hat{p}\hat{q}}{n}} = 0.459 \pm z_{0.10}\sqrt{\dfrac{(0.459)(0.541)}{514}}$

          $0.431 < p < 0.487$

    d.   The 95% confidence interval is wider than the 80% confidence interval. A confidence interval must be wider in order to be more confident that it captures the true value of the population proportion. (See Exercise 4.)

25. No, the confidence interval limits contain the value of 0.13, so the claimed rate of 13% could be the true percentage for the population of brown M&Ms.

    $\hat{p} \pm z_{\alpha/2}\sqrt{\dfrac{\hat{p}\hat{q}}{n}} = 0.08 \pm z_{0.01}\sqrt{\dfrac{(0.08)(0.92)}{100}}$. (Tech: $0.0169 < p < 0.143$)

          $0.0168 < p < 0.143$

27. a. $\hat{p} \pm z_{\alpha/2}\sqrt{\dfrac{\hat{p}\hat{q}}{n}} = 0.000321 \pm z_{0.05}\sqrt{\dfrac{(0.000321)(0.999679)}{420,095}}$ (using $x = 135$: $0.0276\% < p < 0.0367\%$).

          $0.0276\% < p < 0.0366\%$

    b.   No, because 0.0340% is included in the confidence interval.

29. $n = \dfrac{[z_{\alpha/2}]^2\, \hat{p}\hat{q}}{E^2} = \dfrac{[1.645]^2 (0.25)}{0.03^2} = 752$

31. $n = \dfrac{[z_{\alpha/2}]^2\, \hat{p}\hat{q}}{E^2} = \dfrac{[2.575]^2 (0.15)(0.85)}{0.05^2} = 339$

33. a. $n = \dfrac{\left[z_{\alpha/2}\right]^2 \hat{p}\hat{q}}{E^2} = \dfrac{\left[1.96\right]^2 (0.25)}{0.025^2} = 1537$

    b. $n = \dfrac{\left[z_{\alpha/2}\right]^2 \hat{p}\hat{q}}{E^2} = \dfrac{\left[1.96\right]^2 (0.38)(0.62)}{0.025^2} = 1449$

35. a. $n = \dfrac{\left[z_{\alpha/2}\right]^2 \hat{p}\hat{q}}{E^2} = \dfrac{\left[1.645\right]^2 (0.25)}{0.05^2} = 271$

    b. $n = \dfrac{\left[z_{\alpha/2}\right]^2 \hat{p}\hat{q}}{E^2} = \dfrac{\left[1.645\right]^2 (0.85)(0.15)}{0.05^2} = 139$ (Tech: 138)

    c. No. A sample of students at the nearest college is a convenience sample, not a simple random sample, so it is very possible that the results would not be representative of the population of adults.

37. Greater height does not appear to be an advantage for presidential candidates. If greater height is an advantage, then taller candidates should win substantially more than 50% of the elections, but the confidence interval shows that the percentage of elections won by taller candidates is likely to be anywhere between 36.2% and 69.7%.

$$\hat{p} = \frac{18}{34} = 0.529 \,. \qquad \hat{p} \pm z_{\alpha/2}\sqrt{\frac{\hat{p}\hat{q}}{n}} = \frac{18}{34} \pm 1.96\sqrt{\frac{\left(\frac{18}{34}\right)\left(\frac{16}{34}\right)}{34}}$$
$$0.362 < p < 0.697 \text{ or } 36.2\% < p < 69.7\%.$$

39. a. $n = \dfrac{N\hat{p}\hat{q}\left[z_{\alpha/2}\right]^2}{\hat{p}\hat{q}\left[z_{\alpha/2}\right]^2 + (N-1)E^2} = \dfrac{200(0.5)(0.5)\left[1.96\right]^2}{(0.5)(0.5)\left[1.96\right]^2 + (200-1)0.025^2} = 178$

    b. $n = \dfrac{N\hat{p}\hat{q}\left[z_{\alpha/2}\right]^2}{\hat{p}\hat{q}\left[z_{\alpha/2}\right]^2 + (N-1)E^2} = \dfrac{200(0.38)(0.62)\left[1.96\right]^2}{(0.38)(0.62)\left[1.96\right]^2 + (200-1)0.025^2} = 176$

41. The upper confidence interval limit is greater than 100%. Given that the percentage cannot exceed 100%, change the upper limit to 100%.

$$\hat{p} \pm z_{\alpha/2}\sqrt{\frac{\hat{p}\hat{q}}{n}} = \frac{44}{48} \pm 2.575\sqrt{\frac{\left(\frac{44}{48}\right)\left(\frac{4}{48}\right)}{48}}$$
$$0.814 < p < 1.019 \text{ or } 81.4\% < p < 101.9\%.$$

43. Because we have 95% confidence that p is greater than 0.831, we can safely conclude that more than 75% of adults know what Twitter is.

$$\hat{p} + z_{\alpha}\sqrt{\frac{\hat{p}\hat{q}}{n}} = \frac{44}{48} + 1.645\sqrt{\frac{(0.85)(0.15)}{1007}} \quad \text{(Tech: } p > 0.832\text{)}.$$
$$p > 0.831$$

Section 7-3

1. a. $\sec 233.4 \text{ sec} < \mu < 256.65 \text{ sec}$

    b. The best point estimate of μ is $\bar{x} = \dfrac{256.65 + 233.4}{2} = 245.025 \text{ sec}$. The margin of error is

    $E = \dfrac{256.65 - 233.4}{2} = 11.625 \text{ sec.}$

3. We have 95% confidence that the limits of 233.4 sec and 256.65 sec contain the true value of the mean of the population of all duration times.

5. Neither the normal nor the Student t distribution applies.

7. $t_{\alpha/2} = 2.708$

9. Because the sample size is greater than 30, the confidence interval yields a reasonable estimate of μ, even though the data appear to be from a population that is not normally distributed.

$$\bar{x} \pm t_{\alpha/2} \frac{s}{\sqrt{n}} = 9.808 \pm 2.403 \cdot \frac{5.013}{\sqrt{50}}$$

$$8.104 \text{ km} < \mu < 11.512 \text{ km}$$

(Tech: 8.103 km $< \mu <$ 513 km)

11. The $1 salary of Jobs is an outlier that is very far away from the other values, and that outlier has a dramatic effect on the confidence interval.

$$\bar{x} \pm t_{\alpha/2} \frac{s}{\sqrt{n}} = 12898 \pm 2.776 \cdot \frac{7719.05}{\sqrt{5}}$$

$$3315.1 \text{ thousand dollars} < \mu < 22480.9 \text{ thousand dollars}$$

(Tech: 3313.5 thousand dollars $< \mu <$ 6 22,482.5 thousand dollars)

13. Because the confidence interval does not contain 98.6°F, it appears that the mean body temperature is not 98.6°F, as is commonly believed.

$$\bar{x} \pm t_{\alpha/2} \frac{s}{\sqrt{n}} = 98.2 \pm 1.98 \cdot \frac{0.62}{\sqrt{106}}$$

$$98.08°\,\text{F} < \mu < 98.32°\,\text{F}$$

15. Because the confidence interval includes the value of 0, it is very possible that the mean of the changes in LDL cholesterol is equal to 0, suggesting that the garlic treatment did not affect LDL cholesterol levels. It does not appear that garlic is effective in reducing LDL cholesterol.

$$\bar{x} \pm t_{\alpha/2} \frac{s}{\sqrt{n}} = 0.4 \pm 2.4 \cdot \frac{21}{\sqrt{49}}$$

$$-6.8 \text{ mg/dL} < \mu < 7.6 \text{ mg/dL}$$

17. The data appear to have a distribution that is far from normal, so the confidence interval might not be a good estimate of the population mean. The population is likely to be the list of box office receipts for each day of the movie's release. Because the values are from the first 14 days of release, the sample values are not a simple random sample, and they are likely to be the largest of all such values, so the confidence interval is not a good estimate of the population mean.

$$\bar{x} \pm t_{\alpha/2} \frac{s}{\sqrt{n}} = 16.4 \pm 3.01 \cdot \frac{14.5}{\sqrt{14}}$$

$$4.7 \text{ million dollars} < \mu < 28.1 \text{ million dollars}$$

19. The sample data meet the loose requirement of having a normal distribution. Because the confidence interval is entirely below the standard of 1.6 W/kg, it appears that the mean amount of cell phone radiation is less than the FCC standard, but there could be individual cell phones that exceed the standard.

$$\bar{x} \pm t_{\alpha/2} \frac{s}{\sqrt{n}} = 0.938 \pm 1.81 \cdot \frac{0.423}{\sqrt{11}}$$

$$0.707 \text{ W/kg} < \mu < 1.169 \text{ W/kg}$$

21. The sample data meet the loose requirement of having a normal distribution. We cannot conclude that the population mean is less than 7 $\mu g/g$, because the confidence interval shows that the mean might be greater than that level.

$$\bar{x} \pm t_{\alpha/2} \frac{s}{\sqrt{n}} = 11.05 \pm 2.26 \cdot \frac{6.46}{\sqrt{10}}$$

$$6.43 \ \mu g/g < \mu < 15.67 \ \mu g/g$$

23. Although final conclusions about means of populations should not be based on the overlapping of confidence intervals, the confidence intervals do overlap, so it appears that both populations could have the same mean, and there is not clear evidence of discrimination based on age.

CI for ages of unsuccessful applicants

$$\bar{x} \pm t_{\alpha/2} \frac{s}{\sqrt{n}} = 46.96 \pm 2.07 \cdot \frac{7.2}{\sqrt{23}}$$

43.9 years $< \mu < 50.1$ years

CI for ages of successful applicants

$$\bar{x} \pm t_{\alpha/2} \frac{s}{\sqrt{n}} = 44.5 \pm 2.05 \cdot \frac{5.03}{\sqrt{30}}$$

42.6 years $< \mu < 46.4$ years

25. The sample size is $n = \left[\frac{z_{\alpha/2}\sigma}{E} \right]^2 = \left[\frac{1.645 \cdot 15}{3} \right]^2 = 68$, and it does appear to be very reasonable.

27. The required sample size is $n = \left[\frac{z_{\alpha/2}\sigma}{E} \right]^2 = \left[\frac{2.33 \cdot 2157}{250} \right]^2 = 405$ (Tech: 403). It is not likely that you would find that many two-year-old used Corvettes in your region.

29. Use $\sigma = \frac{2400 - 600}{4} = 450$ to get a sample size of $n = \left[\frac{z_{\alpha/2}\sigma}{E} \right]^2 = \left[\frac{2.33 \cdot 450}{100} \right]^2 = 110$. The margin of error of 100 points seems too high to provide a good estimate of the mean SAT score.

31. With the range rule of thumb, use $\sigma = \frac{90 - 46}{4} = 11$ to get a required sample size of

$$n = \left[\frac{z_{\alpha/2}\sigma}{E} \right]^2 = \left[\frac{1.96 \cdot 11}{2} \right]^2 = 117 \ .$$

With $s = 10.3$, the required sample size is $n = \left[\frac{z_{\alpha/2}\sigma}{E} \right]^2 = \left[\frac{1.96 \cdot 10.3}{2} \right]^2 = 102$. The better estimate of s is the standard deviation of the sample, so the correct sample size is likely to be closer to 102 than 117.

33. $\bar{x} \pm t_{\alpha/2} \frac{s}{\sqrt{n}} = 1.1842 \pm 2.8 \cdot \frac{0.5873}{\sqrt{50}}$

$0.963 < \mu < 1.407$

(Tech: $0.962 < \mu < 1.407$)

35. $\bar{x} \pm z_{\alpha/2} \frac{\sigma}{\sqrt{n}} = 9.808 \pm 2.33 \cdot \frac{5.013}{\sqrt{50}}$

8.156 km $< \mu < 11.46$ km

(Tech: 8.159 km $< \mu < 11.457$ km)

37. $$\bar{x} \pm z_{\alpha/2} \frac{\sigma}{\sqrt{n}} = 12898 \pm 1.96 \cdot \frac{7717.8}{\sqrt{5}}$$

6133.05 thousand dollars $< \mu < 19662.95$ thousand dollars

(Tech: 6131.9 thousand dollars $< \mu < 19{,}663.3$ thousand dollars)

39. The sample data do not appear to meet the loose requirement of having a normal distribution. The effect of the outlier on the confidence interval is very substantial. Outliers should be discarded if they are known to be errors. If an outlier is a correct value, it might be very helpful to see its effects by constructing the confidence interval with and without the outlier included.

$$\bar{x} \pm t_{\alpha/2} \frac{s}{\sqrt{n}} = 11.375 \pm 2.26 \cdot \frac{7.1851}{\sqrt{10}}$$

$$-24.54 \text{ m} < \mu < 106.04 \text{ m}$$

(Tech: $-24.55 \text{ m} < \mu < 106.05 \text{ m}$)

41. The confidence interval based on the first sample value is much wider than the confidence interval based on all 10 sample values.

$$x \pm 9.68|3.0|$$

$$-26.0 \text{ m} < \mu < 32.0 \text{ m}$$

Section 7-4

1. $\sqrt{916.591 \ (\text{mg/dL})^2} < \sqrt{\sigma^2} < \sqrt{2252.1149 \ (\text{mg/dL})^2} \Rightarrow 30.3 \text{ mg/dL} < \sigma < 47.5 \text{ mg/dL}$. We have 95% confidence that the limits of 30.3 mg/dL and 47.5 mg/dL contain the true value of the standard deviation of the LDL cholesterol levels of all women.

3. The original sample values can be identified, but the dotplot shows that the sample appears to be from a population having a uniform distribution, not a normal distribution as required. Because the normality requirement is not satisfied, the confidence interval estimate of s should not be constructed using the methods of this section.

5. df = 24. $\chi_L^2 = 9.886$ and $\chi_R^2 = 45.559$.

$$\sqrt{\frac{(n-1)s^2}{\chi_R^2}} < \sigma < \sqrt{\frac{(n-1)s^2}{\chi_L^2}}$$

$$\sqrt{\frac{(25-1)0.24^2}{45.559}} < \sigma < \sqrt{\frac{(25-1)0.24^2}{9.886}}$$

$$0.17 \text{ mg} < \sigma < 0.37 \text{ mg}$$

7. df = 39. $\chi_L^2 = 24.433$ (Tech: 23.654) and $\chi_R^2 = 59.342$ (Tech: 58.120).

$$\sqrt{\frac{(n-1)s^2}{\chi_R^2}} < \sigma < \sqrt{\frac{(n-1)s^2}{\chi_L^2}}$$

$$\sqrt{\frac{(40-1)65.2^2}{59.342}} < \sigma < \sqrt{\frac{(40-1)65.2^2}{24.433}}; \text{df} = 40$$

$$52.9 < \sigma < 82.4 \text{ (Tech: } 53.4 < \sigma < 83.7)$$

9.
$$\sqrt{\frac{(n-1)s^2}{\chi_R^2}} < \sigma < \sqrt{\frac{(n-1)s^2}{\chi_L^2}}$$

$$\sqrt{\frac{(106-1)0.62^2}{124.342}} < \sigma < \sqrt{\frac{(106-1)0.62^2}{77.929}}; \text{df} = 100$$

$$0.579°\text{F} < \sigma < 0.720°\text{F} \text{ (Tech: } 0.557°\text{F 6 s 6 } 0.700°\text{F)}$$

11. The confidence interval shows that the standard deviation is not likely to be less than 30 mL, so the variation is too high instead of being at an acceptable level below 30 mL. (Such one-sided claims should be tested using the formal methods presented in Chapter 8.)

$$\sqrt{\frac{(n-1)s^2}{\chi_R^2}} < \sigma < \sqrt{\frac{(n-1)s^2}{\chi_L^2}}$$

$$\sqrt{\frac{(24-1)42.8^2}{44.181}} < \sigma < \sqrt{\frac{(24-1)42.8^2}{9.260}}$$

$$30.9 \text{ mL} < \sigma < 67.45 \text{ mL}$$

13.
$$\sqrt{\frac{(n-1)s^2}{\chi_R^2}} < \sigma < \sqrt{\frac{(n-1)s^2}{\chi_L^2}}$$

$$\sqrt{\frac{(7-1)0.36576^2}{12.592}} < \sigma < \sqrt{\frac{(7-1)0.36576^2}{1.635}} \, 0$$

$$0.252 \text{ ppm} < \sigma < 0.701 \text{ ppm}$$

15. CI for ages of unsuccessful applicants:

$$\sqrt{\frac{(n-1)s^2}{\chi_R^2}} < \sigma < \sqrt{\frac{(n-1)s^2}{\chi_L^2}}$$

$$\sqrt{\frac{(25-1)7.22^2}{45.559}} < \sigma < \sqrt{\frac{(25-1)7.22^2}{9.886}}$$

$$5.2 \text{ years} < \sigma < 11.5 \text{ years}$$

CI for ages of successful applicants:

$$\sqrt{\frac{(n-1)s^2}{\chi_R^2}} < \sigma < \sqrt{\frac{(n-1)s^2}{\chi_L^2}}$$

$$\sqrt{\frac{(29-1)5.026^2}{50.993}} < \sigma < \sqrt{\frac{(29-1)5.026^2}{12.461}}$$

$$3.7 \text{ years} < \sigma < 7.5 \text{ years}$$

Although final conclusions about means of populations should not be based on the overlapping of confidence intervals, the confidence intervals do overlap, so it appears that the two populations have standard deviations that are not dramatically different.

17.
$$\sqrt{\frac{(n-1)s^2}{\chi_R^2}} < \sigma < \sqrt{\frac{(n-1)s^2}{\chi_L^2}}$$

$$\sqrt{\frac{(37-1)0.0165^2}{63.691}} < \sigma < \sqrt{\frac{(37-1)0.0165^2}{22.164}}; \text{ df=40}$$

$$0.01239 \text{ g} < \sigma < 0.02111 \text{ g}$$

(Tech: $0.01291 \text{ g} < \sigma < 0.02255 \text{ g}$)

19. 33,218 is too large. There aren't 33,218 statistics professors in the population, and even if there were, that sample size is too large to be practical.

21. The sample size is 768. Because the population does not have a normal distribution, the computed minimum sample size is not likely to be correct.

23. $\chi_L^2 = \frac{1}{2}\left[-z_{\alpha/2} + \sqrt{2k-1}\right]^2 = \frac{1}{2}\left[-1.645 + \sqrt{2\cdot105-1}\right]^2 = 82.072$ and

$\chi_R^2 = \frac{1}{2}\left[z_{\alpha/2} + \sqrt{2k-1}\right]^2 = \frac{1}{2}\left[1.645 + \sqrt{2\cdot105-1}\right]^2 = 129.635$

(Tech using $z_{\alpha/2} = 1.644853626$: $\chi_L^2 = 82.073$ and $\chi_R^2 = 129.632$). The approximate values are quite close to the actual critical values.

Chapter Quick Quiz

1. $40\% - 3.1\% < p < 40\% + 3.1\%$

   $36.9\% < p < 43.1\%$

2. $\hat{p} = \dfrac{0.511 + 0.449}{2} = 0.480$

3. We have 95% confidence that the limits of 0.449 and 0.511 contain the true value of the proportion of females in the population of medical school students.

4. $z = 1.645$

5. $n = \dfrac{[z_{\alpha/2}]^2 \,\hat{p}\hat{q}}{E^2} = \dfrac{[1.645]^2\,(0.25)}{0.03^2} = 752$

6. $n = \left[\dfrac{z_{\alpha/2}\sigma}{E}\right]^2 = \left[\dfrac{2.575\cdot15}{2}\right]^2 = 373$ (Tech: 374)

7. The sample must be a simple random sample and there is a loose requirement that the sample values appear to be from a normally distributed population.

8. The degrees of freedom is the number of sample values that can vary after restrictions have been imposed on all of the values. For the sample data in Exercise 7, df = 5.

9. $t = 2.571$

10. $\chi_L^2 = 0.831$ and $\chi_R^2 = 12.833$

Review Exercises

1. a. $\hat{p} = \dfrac{284}{557} = 0.510 = 51.0\%$

   b. $\hat{p} \pm z_{\alpha/2}\sqrt{\dfrac{\hat{p}\hat{q}}{n}} = \dfrac{284}{557} \pm 1.96\sqrt{\dfrac{\left(\frac{284}{557}\right)\left(\frac{273}{557}\right)}{557}}$

   $46.8\% < p < 55.1\%$

   c. No, the confidence interval shows that the population percentage might be 50% or less, so we cannot safely conclude that the majority of adults say that they are underpaid.

2. $n = \dfrac{[z_{\alpha/2}]^2\,\hat{p}\hat{q}}{E^2} = \dfrac{[2.575]^2\,(0.25)}{0.02^2} = 4145$ (Tech: 4147)

3. $n = \left[\dfrac{z_{\alpha/2}\sigma}{E}\right]^2 = \left[\dfrac{2.33\cdot16}{3}\right]^2 = 155$ (Tech: 154)

4. a.   Student t distribution

    b.   Normal distribution

    c.   The distribution is not normal, Student t, or chi-square.

    d. χ^2 (chi-square distribution)

    e.   Normal distribution

5. a. $n = \dfrac{\left[z_{\alpha/2}\right]^2 \hat{p}\hat{q}}{E^2} = \dfrac{[2.33]^2 (0.25)}{0.05^2} = 543$ (Tech: 542)

    b. $n = \left[\dfrac{z_{\alpha/2}\sigma}{E}\right]^2 = \left[\dfrac{2.33 \cdot 337}{50}\right]^2 = 247$ (Tech: 246)

    c.   543

6. Because the entire confidence interval is above 50%, we can safely conclude that the majority of adults consume alcoholic beverages.

$$\hat{p} \pm z_{\alpha/2}\sqrt{\dfrac{\hat{p}\hat{q}}{n}} = 0.64 \pm 1.65\sqrt{\dfrac{(0.64)(0.36)}{1011}}$$
$$61.5\% < p < 66.5\%$$

7. $\bar{x} \pm t_{\alpha/2}\dfrac{s}{\sqrt{n}} = 143 \pm 2.201 \cdot \dfrac{259.7754}{\sqrt{12}}$

    $-22.1 \text{ sec} < \mu < 308.1 \text{ sec}$

8. Because women and men have some notable physiological differences, the confidence interval does not necessarily serve as an estimate of the mean white blood cell count of men.

$$\bar{x} \pm t_{\alpha/2}\dfrac{s}{\sqrt{n}} = 7.15 \pm 1.685 \cdot \dfrac{2.28}{\sqrt{40}}$$
$$6.54 < \mu < 7.76$$

9. There is 95% confidence that the limits of 37.5 g and 47.9 g contain the true mean deceleration measurement for all small cars.

$$\bar{x} \pm t_{\alpha/2}\dfrac{s}{\sqrt{n}} = 42.7 \pm 2.447 \cdot \dfrac{5.6}{\sqrt{7}}$$
$$37.5 \text{ g} < \mu < 47.9 \text{ g}$$

10. $\sqrt{\dfrac{(n-1)s^2}{\chi_R^2}} < \sigma < \sqrt{\dfrac{(n-1)s^2}{\chi_L^2}}$

$\sqrt{\dfrac{(7-1)5.6^2}{14.449}} < \sigma < \sqrt{\dfrac{(7-1)5.6^2}{1.237}}$

$3.6 \text{ g} < \sigma < 12.3 \text{ g}$

Cumulative Review Exercises

1. $\bar{x} = 5.5$; median = 5.0; $s = 3.8$

2. The range of usual values is from $5.5 - 2(3.8) = -2.1$ to $5.5 + 2(3.8) = 3.8$ (or from 0 to 13.1).

3. Ratio level of measurement; discrete data.

4. $n = \left[\dfrac{z_{\alpha/2}\sigma}{E}\right]^2 = \left[\dfrac{1.96 \cdot 5.8}{2}\right]^2 = 33$ campuses

5. The population should include only colleges of the same type as the sample, so the population consists of all large urban campuses with residence halls.

$$\bar{x} \pm t_{\alpha/2} \frac{s}{\sqrt{n}} = 5.5 \pm 2.02 \cdot \frac{5.8}{\sqrt{40}}$$

$$3.6 < \mu < 7.4$$

6. The graphs suggest that the population has a distribution that is skewed (to the right) instead of being normal. The histogram shows that some taxi-out times can be very long, and that can occur with heavy traffic, but little or no traffic cannot make the taxi-out time very low. There is a minimum time required, regardless of traffic conditions. Construction of a confidence interval estimate of a population standard deviation has a strict requirement that the sample data are from a normally distributed population, and the graphs show that this strict normality requirement is not satisfied.

7. a. $$\hat{p} \pm z_{\alpha/2} \sqrt{\frac{\hat{p}\hat{q}}{n}} = 0.59 \pm 1.96 \sqrt{\frac{(0.59)(0.31)}{1003}}$$ (or $0.560 < p < 0.621$ if using $x = 592$)

    $$0.560 < p < 0.620$$

    b.   Because the survey was about shaking hands and because it was sponsored by a supplier of hand sanitizer products, the sponsor could potentially benefit from the results, so there might be some pressure to obtain results favorable to the sponsor.

    c. $$n = \frac{\left[z_{\alpha/2}\right]^2 \hat{p}\hat{q}}{E^2} = \frac{\left[1.96\right]^2 (0.25)}{0.025^2} = 1083$$

8. There does not appear to be a correlation between HDL and LDL cholesterol levels.

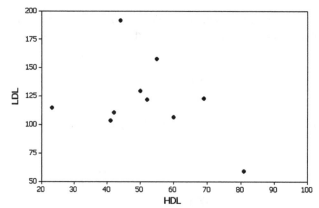

9. a. $$z = \frac{185 - 175}{9} = 1.11 \text{ and } P(z > 1.11) = 13.35\% \text{ (Tech: 13.32\%)}.$$

    Yes, losing about 13% of the market would be a big loss.

    b.   5th percentile: $x = \mu + z \cdot \sigma = 175 - 1.645 \cdot 9 = 160.2$ mm

    95th percentile: $x = \mu + z \cdot \sigma = 175 + 1.645 \cdot 9 = 189.8$ mm

10. a.   There are 10^3 possible tickets so the probability of winning by purchasing one ticket is $\dfrac{1}{1000}$.

    b. $1 - \dfrac{1}{1000} = \dfrac{999}{1000}.$ c. $\left(\dfrac{999}{1000}\right)^{10} = 0.990.$

Chapter 8: Hypothesis Testing

Section 8-2

1. Rejection of the aspirin claim is more serious because the aspirin is a drug treatment. The wrong aspirin dosage can cause adverse reactions. M&Ms do not have those same adverse reactions. It would be wise to use a smaller significance level for testing the aspirin claim.

3. a. H_0: $\mu = 98.6°F$

    b. H_1: $\mu \neq 98.6°F$

    c.   Reject the null hypothesis or fail to reject the null hypothesis.

    d.   No. In this case, the original claim becomes the null hypothesis. For the claim that the mean body temperature is equal to 98.6°F, we can either reject that claim or fail to reject it, but we cannot state that there is sufficient evidence to support that claim.

5. a. $p = 0.20$

    b. H_0: $p = 0.20$ and H_1: $p \neq 0.20$

7. a. $\mu \leq 76$

    b. H_0: $\mu = 76$ and H_1: $\mu < 76$

9. There is not sufficient evidence to warrant rejection of the claim that 20% of adults smoke.

11. There is not sufficient evidence to warrant rejection of the claim that the mean pulse rate of adult females is 76 or lower.

13. $z = \dfrac{\hat{p} - p}{\sqrt{\dfrac{pq}{n}}} = \dfrac{0.89 - 0.75}{\sqrt{\dfrac{(0.75)(0.25)}{1021}}} = 10.33$ (or $z = 10.35$ if using $x = 909$)

15. $\chi^2 = \dfrac{(n-1)s^2}{\sigma^2} = \dfrac{(40-1)2.28^2}{5^2} = 8.110$

17. P-value $= P(z > 2) = 0.0228$. Critical value: $z = 1.645$.

19. P-value $= 2 \cdot P(z < -1.75) = 0.0802$. (Tech: 0.0801). Critical values: $z = -1.96$, $z = 1.96$.

21. P-value $= 2 \cdot P(z < -1.23) = 0.2186$. (Tech: 0.2187). Critical values: $z = -1.96$, $z = 1.96$.

23. P-value $= P(z < -3.00) = 0.0013$. Critical value: $z = -1.645$.

25. a.   Reject H_0.

    b.   There is sufficient evidence to support the claim that the percentage of blue M&Ms is greater than 5%.

27. a.   Fail to reject H_0.

    b.   There is not sufficient evidence to warrant rejection of the claim that women have heights with a mean equal to 160.00 cm.

29. a. H_0: $p = 0.5$ and H_1: $p > 0.5$ e. $z = 1.00$

    b. $\alpha = 0.01$ f. P-value $= P(z > 1.00) = 0.1587$.

    c.   Normal distribution.   g. $z = 2.33$

    d.   Right-tailed.   h.   0.01

31. Type I error: In reality $p = 0.1$, but we reject the claim that $p = 0.1$. Type II error: In reality $p \neq 0.1$, but we fail to reject the claim that $p = 0.1$.

33. Type I error: In reality $p = 0.5$, but we support the claim that $p > 0.5$. Type II error: In reality $p > 0.5$, but we fail to support that conclusion.

35. The power of 0.96 shows that there is a 96% chance of rejecting the null hypothesis of $p = 0.08$ when the true proportion is actually 0.18. That is, if the proportion of Chantix users who experience abdominal pain is actually 0.18, then there is a 96% chance of supporting the claim that the proportion of Chantix users who experience abdominal pain is greater than 0.08.

37. From $p = 0.5$, $\hat{p} = 0.5 + 1.645\sqrt{\dfrac{(0.5)(0.5)}{n}}$, from $p = 0.55$, $\hat{p} = 0.55 - 0.842\sqrt{\dfrac{(0.55)(0.45)}{n}}$; Since $\left(P(z > -0.842) = 0.8000\right)$, so:

$$0.5 + 1.645\sqrt{\frac{(0.5)(0.5)}{n}} = 0.55 - 0.842\sqrt{\frac{(0.55)(0.45)}{n}}$$
$$0.5\sqrt{n} + 1.645\sqrt{0.25} = 0.55\sqrt{n} - 0.842\sqrt{0.2475}$$
$$0.05\sqrt{n} = 1.645\sqrt{0.25} + 0.842\sqrt{0.2475}$$
$$n = \left(\frac{1.645\sqrt{0.25} + 0.842\sqrt{0.2475}}{0.05}\right)^2 = 617$$

Section 8-3

1. The *P*-value method and the critical value method always yield the same conclusion. The confidence interval method might or might not yield the same conclusion obtained by using the other two methods.

3. *P*-value = 0.00000000550. Because the *P*-value is so low, we have sufficient evidence to support the claim that $p < 0.5$.

5. a. Left-tailed.

    b. $z = -1.94$

    c. *P*-value = 0.0260 (rounded)

    d. H_0: $p = 0.1$. Reject the null hypothesis.

    e. There is sufficient evidence to support the claim that less than 10% of treated subjects experience headaches.

7. a. Two-tailed.

    b. $z = -0.82$

    c. *P*-value = 0.4106

    d. H_0: $p = 0.35$. Fail to reject the null hypothesis.

    e. There is not sufficient evidence to warrant rejection of the claim that 35% of adults have heard of the Sony Reader.

9. H_0: $p = 0.25$. H_1: $p \neq 0.25$. Test statistic: $z = \dfrac{\frac{152}{580} - 0.25}{\sqrt{\frac{(0.25)(0.75)}{580}}} = 0.67$. Critical values: $z = \pm 2.575$

(Tech: ± 2.576). *P*-value $= 2 \cdot P(z > 0.67) = 0.5028$ (Tech: 0.5021). Fail to reject H_0. There is not sufficient evidence to warrant rejection of the claim that 25% of offspring peas will be yellow.

```
MINITAB
Test of p = 0.25 vs p not = 0.25
Sample    X     N Sample p        95% CI          Z-Value  P-Value
1        152   580 0.262069  (0.226280, 0.297858)    0.67    0.502
```

11. H_0: $p = 0.5$. H_1: $p > 0.5$. Test statistic: $z = \dfrac{\frac{531}{1002} - 0.5}{\sqrt{\frac{(0.5)(0.5)}{1002}}} = 1.90$. Critical value: $z = 1.645$. P-value

$= P(z > 1.90) = 0.0287$ (Tech: 0.0290). Reject H_0. There is sufficient evidence to support the claim that the majority of adults feel vulnerable to identify theft.

```
MINITAB
Test of p = 0.5 vs p > 0.5
Sample    X      N    Sample p    Z-Value    P-Value
1        531    1002   0.529940     1.90      0.029
```

13. H_0: $p = 0.5$. H_1: $p > 0.5$. Test statistic: $z = \dfrac{\frac{879}{945} - 0.5}{\sqrt{\frac{(0.5)(0.5)}{945}}} = 26.45$. Critical value: $z = 2.33$. P-value

$= P(z > 26.45) = 0.0001$ (Tech: 0.0000). Reject H_0. There is sufficient evidence to support the claim that the XSORT method is effective in increasing the likelihood that a baby will be a girl.

```
MINITAB
Test of p = 0.5 vs p > 0.5
Sample    X      N    Sample p    Z-Value    P-Value
1        879    945    0.930159    26.45      0.000
```

15. H_0: $p = 0.5$. H_1: $p \neq 0.5$. Test statistic: $z = \dfrac{\frac{123}{280} - 0.5}{\sqrt{\frac{(0.5)(0.5)}{280}}} = -2.03$. Critical values: $z = \pm 1.645$. P-value

$= 2 \cdot P(z < -2.03) = 0.0424$ (Tech: 0.0422). Reject H_0. There is sufficient evidence to warrant rejection of the claim that touch therapists use a method equivalent to random guesses. However, their success rate of 123/280 (or 43.9%) indicates that they performed worse than random guesses, so they do not appear to be effective.

```
MINITAB
Test of p = 0.5 vs p not = 0.5
Sample    X      N    Sample p        95% CI           Z-Value    P-Value
1        123    280    0.439286   (0.381154, 0.497417)   -2.03     0.042
```

17. H_0: $p = \frac{1}{3}$. H_1: $p < \frac{1}{3}$. Test statistic: $z = \dfrac{\frac{172}{611} - \frac{1}{3}}{\sqrt{\frac{\left(\frac{1}{3}\right)\left(\frac{2}{3}\right)}{611}}} = -2.72$. Critical value: $z = -2.33$. P-value

$= P(z < -2.72) = 0.0033$. Reject H_0. There is sufficient evidence to support the claim that fewer than 1/3 of the challenges are successful. Players don't appear to be very good at recognizing referee errors.

```
MINITAB
Test of p = 0.3333 vs p < 0.3333
Sample    X      N    Sample p    Z-Value    P-Value
1        172    611    0.281506    -2.72      0.003
```

19. H_0: $p = 0.000340$. $p \neq 0.000340$. Test statistic: $z = \dfrac{\frac{135}{420,095} - 0.000340}{\sqrt{\frac{(0.000340)(0.99966)}{420,095}}} = -0.66$. Critical values:

$z = \pm 2.81$. P-value $= 2 \cdot P(z < -0.66) = 0.5092$ (Tech: 0.5122). Fail to reject H_0. There is not sufficient evidence to support the claim that the rate is different from 0.0340%. Cell phone users should not be concerned about cancer of the brain or nervous system.

```
MINITAB
Test of p = 0.00034 vs p not = 0.00034
Sample    X       N     Sample p        95% CI           Z-Value    P-Value
1        135   420095   0.000321   (0.000267, 0.000376)   -0.66     0.512
```

21. H_0: $p = 0.5$. H_1: $p \neq 0.5$. Test statistic: $z = \dfrac{\frac{235}{414} - 0.5}{\sqrt{\frac{(0.5)(0.5)}{414}}} = 2.75$. Critical values: $z = \pm 1.96$. P-value

$= 2 \cdot P(z > 2.75) = 0.0060$ (Tech: 0.0059). Reject H_0. There is sufficient evidence to warrant rejection of the claim that the coin toss is fair in the sense that neither team has an advantage by winning it. The coin toss rule does not appear to be fair.

MINITAB
Test of p = 0.5 vs p not = 0.5

Sample	X	N	Sample p	95% CI	Z-Value	P-Value
1	235	414	0.567633	(0.519912, 0.615354)	2.75	0.006

23. H_0: $p = 0.5$. H_1: $p < 0.5$. Test statistic: $z = \dfrac{\frac{152}{380} - 0.5}{\sqrt{\frac{(0.5)(0.5)}{380}}} = -3.90$. Critical value: $z = -2.33$. P-value

$= P(z < -3.90) = 0.0001$ (Tech: 0.0000484). Reject H_0. There is sufficient evidence to support the claim that fewer than half of smartphone users identify the smartphone as the only thing they could not live without. Because only smartphone users were surveyed, the results do not apply to the general population.

MINITAB
Test of p = 0.5 vs p < 0.5

Sample	X	N	Sample p	Z-Value	P-Value
1	152	380	0.400000	-3.90	0.000

25. H_0: $p = 0.25$. H_1: $p > 0.25$. Test statistic: $z = \dfrac{0.29 - 0.25}{\sqrt{\frac{(0.25)(0.75)}{427}}} = 1.91$ or $z = 1.93$ (using x = 124). Critical

value: $z = 1.645$ (assuming a 0.05 significance level). P-value $= P(z > 1.92) = 0.0281$ (using $\hat{p} = 0.29$) or 0.0268 (using $x = 124$) (Tech P-value = 0.0269). Reject H_0. There is sufficient evidence to support the claim that more than 25% of women purchase books online.

MINITAB
Test of p = 0.25 vs p > 0.25

Sample	X	N	Sample p	Z-Value	P-Value
1	124	427	0.290398	1.93	0.027

27. H_0: $p = 0.75$. H_1: $p > 0.75$. Test statistic: $z = \dfrac{0.90 - 0.75}{\sqrt{\frac{(0.75)(0.25)}{514}}} = 7.85$ or $z = 7.89$ (using x = 463). Critical

value: $z = 2.33$. P-value $= P(z > 7.85) = 0.0001$ (Tech: 0.0000). Reject H_0. There is sufficient evidence to support the claim that more than 3/4 of all human resource professionals say that the appearance of a job applicant is most important for a good first impression.

MINITAB
Test of p = 0.75 vs p > 0.75

Sample	X	N	Sample p	Z-Value	P-Value
1	463	514	0.900778	7.89	0.000

29. H_0: $p = 0.791$. H_1: $p < 0.791$. Test statistic: $z = \dfrac{0.39 - 0.791}{\sqrt{\frac{(0.791)(0.209)}{870}}} = -29.09$ or $z = -29.11$ (using x = 339).

Critical value: $z = -2.33$. P-value $= 2 \cdot P(z < -29.09) = 0.0001$ (Tech: 0.0000). Reject H_0. There is sufficient evidence to support the claim that the percentage of selected Americans of Mexican ancestry is less than 79.1%, so the jury selection process appears to be unfair.

MINITAB
Test of p = 0.791 vs p < 0.791

Sample	X	N	Sample p	Z-Value	P-Value
1	339	870	0.389655	-29.11	0.000

31. H_0: $p = 0.75$. H_1: $p > 0.75$. Test statistic: $z = \dfrac{0.77 - 0.75}{\sqrt{\frac{(0.75)(0.25)}{25,000}}} = 7.30$. Critical value: $z = 2.33$. P-value

$= P(z > 7.30) = 0.0001$ (Tech: 0.0000). Reject H_0. There is sufficient evidence to support the claim that more than 75% of television sets in use were tuned to the Super Bowl.

MINITAB
Test of p = 0.75 vs p > 0.75

Sample	X	N	Sample p	Z-Value	P-Value
1	19250	25000	0.770000	7.30	0.000

33. Among 100 M&Ms, 19 are green. H_0: $p = 0.16$. H_1: $p \neq 0.16$. Test statistic: $z = \dfrac{0.19 - 0.16}{\sqrt{\frac{(0.16)(0.84)}{100}}} = 0.82$.

Critical values: $z = \pm 1.96$. P-value $= 2 \cdot P(z > 0.82) = 0.4122$ (Tech: 0.4132). Fail to reject H_0. There is not sufficient evidence to warrant rejection of the claim that 16% of plain M&M candies are green.

MINITAB
Test of p = 0.16 vs p not = 0.16

Sample	X	N	Sample p	95% CI	Z-Value	P-Value
1	19	100	0.190000	(0.113110, 0.266890)	0.82	0.413

35. H_0: $p = 0.5$. H_1: $p > 0.5$. Using the binomial probability distribution with an assumed proportion of $p = 0.5$, the probability of 7 or more heads is 0.0352, so the P-value is 0.0352. Reject H_0. There is sufficient evidence to support the claim that the coin favors heads.

37. a. From $p = 0.40$, $\hat{p} = 0.4 - 1.645\sqrt{\dfrac{(0.4)(0.6)}{50}} = 0.286$

From $p = 0.25$, $z = \dfrac{0.286 - 0.25}{\sqrt{\frac{(0.25)(0.75)}{50}}} = 0.588$; Power $= P(z < 0.588) = 0.7224$. (Tech: 0.7219)

b. $1 - 0.7224 = 0.2776$ (Tech: 0.2781)

c. The power of 0.7224 shows that there is a reasonably good chance of making the correct decision of rejecting the false null hypothesis. It would be better if the power were even higher, such as greater than 0.8 or 0.9.

Section 8-4

1. The requirements are (1) the sample must be a simple random sample, and (2) either or both of these conditions must be satisfied: The population is normally distributed or $n > 30$. There is not enough information given to determine whether the sample is a simple random sample. Because the sample size is not greater than 30, we must check for normality, but the value of 583 sec appears to be an outlier, and a normal quantile plot or histogram suggests that the sample does not appear to be from a normally distributed population.

1.   (continued)

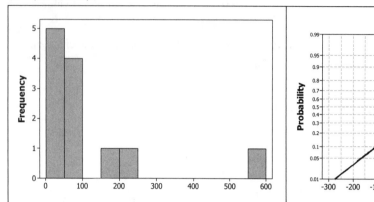 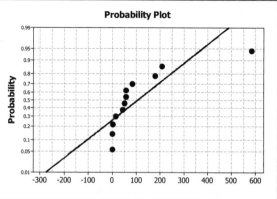

3.   A t test is a hypothesis test that uses the Student t distribution, such as the method of testing a claim about a population mean as presented in this section. The t test methods are much more likely to be used than the z test methods because the t test does not require a known value of σ, and realistic hypothesis tests of claims about μ typically involve a population with an unknown value of σ.

5. P-value < 0.005 (Tech: 0.0013).   7. $0.02 < P$-value < 0.05 (Tech: 0.0365).

9. H_0: $\mu = 24$. H_1: $\mu < 24$. Test statistic: $t = -7.323$. Critical value: $t = -1.685$. P-value < 0.005. (The display shows that the P-value is 0.00000000387325.) Reject H_0. There is sufficient evidence to support the claim that Chips Ahoy reduced-fat cookies have a mean number of chocolate chips that is less than 24 (but this does not provide conclusive evidence of reduced fat).

11. H_0: $\mu = 33$ years. H_1: $\mu \neq 33$ years. Test statistic: $t = \dfrac{35.9 - 33}{11.1/\sqrt{82}} = 2.367$. Critical values $t = \pm 2.639$

(approximately). P-value > 0.02 (Tech: 0.0204). Fail to reject H_0. There is not sufficient evidence to warrant rejection of the claim that the mean age of actresses when they win Oscars is 33 years.

```
MINITAB
Test of mu = 33 vs not = 33
  N   Mean  StDev  SE Mean      95% CI        T      P
  82  35.90  11.10    1.23   (33.46, 38.34)  2.37  0.020
```

13. H_0: $\mu = 0.8535$ g. H_1: $\mu \neq 0.8535$ g. Test statistic: $t = \dfrac{0.8635 - 0.8535}{0.0570/\sqrt{19}} = 0.765$. Critical values:

$t = \pm 2.101$. P-value > 0.20 (Tech: 0.4543). Fail to reject H_0. There is not sufficient evidence to warrant rejection of the claim that the mean weight of all green M&Ms is equal to 0.8535 g. The green M&Ms do appear to have weights consistent with the package label.

```
MINITAB
Test of mu = 0.8535 vs not = 0.8535
  N    Mean   StDev  SE Mean      95% CI         T     P
  19  0.8635  0.0570  0.0131  (0.8360, 0.8910)  0.76  0.454
```

15. H_0: $\mu = 0$ lb. H_1: $\mu > 0$ lb. Test statistic: $t = \dfrac{3 - 0}{4.9/\sqrt{40}} = 3.872$. Critical value: $t = 2.426$.

P-value < 0.005 (Tech: 0.0002). Reject H_0. There is sufficient evidence to support the claim that the mean weight loss is greater than 0. Although the diet appears to have statistical significance, it does not appear to have practical significance, because the mean weight loss of only 3.0 lb does not seem to be worth the effort and cost.

```
MINITAB
Test of mu = 0 vs > 0
  N   Mean   StDev  SE Mean    T      P
  40  3.000  4.900   0.775   3.87  0.000
```

17. H_0: $\mu = 0$. H_1: $\mu > 0$. Test statistic: $t = \dfrac{0.4 - 0}{21.0/\sqrt{49}} = 0.133$. Critical value: $t = 1.676$ (approximately,

assuming a 0.05 significance level). P-value > 0.10 (Tech: 0.4472). Fail to reject H_0. There is not sufficient evidence to support the claim that with garlic treatment, the mean change in LDL cholesterol is greater than 0. The results suggest that the garlic treatment is not effective in reducing LDL cholesterol levels.

MINITAB
Test of mu = 0 vs > 0

N	Mean	StDev	SE Mean	T	P
49	0.40	21.00	3.00	0.13	0.447

19. H_0: $\mu = 4$ years. H_1: $\mu > 4$ years. Test statistic: $t = 3.189$. Critical value: $t = 2.539$.

P-value < 0.005 (Tech: 0.0024). Reject H_0. There is sufficient evidence to support the claim that the mean time required to earn a bachelor's degree is greater than 4.0 years. Because $n \le 30$ and the data do not appear to be from a normally distributed population, the requirement that "the population is normally distributed or $n > 30$" is not satisfied, so the conclusion from the hypothesis test might not be valid. However, some of the sample values are equal to 4 years and others are greater than 4 years, so the claim does appear to be justified.

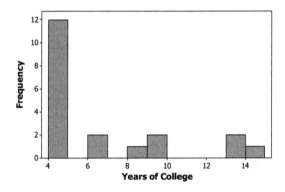

MINITAB
Test of mu = 4 vs > 4

Variable	N	Mean	StDev	SE Mean	T	P
CollegeYears	20	6.500	3.506	0.784	3.19	0.002

21. The sample data meet the loose requirement of having a normal distribution. H_0: $\mu = 14$ mg/g.

H_1: $\mu < 14$ mg/g. Test statistic: $t = -1.444$. Critical value: $t = -1.833$. P-value > 0.05 (Tech: 0.0913).

Fail to reject H_0. There is not sufficient evidence to support the claim that the mean lead concentration for all such medicines is less than 14 mg/g.

MINITAB
Test of mu = 14 vs < 14

Variable	N	Mean	StDev	SE Mean	T	P
Lead	10	11.05	6.46	2.04	-1.44	0.091

23. The sample data meet the loose requirement of having a normal distribution. H_0: $\mu = 63.8$ in.

H_1: $\mu > 63.8$ in. Test statistic: $t = 23.824$. Critical value: $t = 2.821$. P-value < 0.005 (Tech: 0.0000).

Reject H_0. There is sufficient evidence to support the claim that supermodels have heights with a mean that is greater than the mean height of 63.8 in. for women in the general population. We can conclude that supermodels are taller than typical women.

MINITAB
Test of mu = 63.8 vs > 63.8

Variable	N	Mean	StDev	SE Mean	T	P
Heights	10	69.825	0.800	0.253	23.82	0.000

25. The sample data meet the loose requirement of having a normal distribution. H_0: $\mu = 1.00$. H_1: $\mu > 1.00$.
Test statistic: $t = 2.218$. Critical value: $t = 1.676$ (approximately). P-value < 0.025 (Tech: 0.0156). Reject H_0. There is sufficient evidence to support the claim that the population of earthquakes has a mean magnitude greater than 1.00.

```
MINITAB
Test of mu = 1 vs > 1
Variable   N   Mean    StDev   SE Mean   Bound     T       P
MAG       5 0  1.1842  0.5873  0.0831    1.0449    2.22    0.016
```

27. The sample data meet the loose requirement of having a normal distribution. H_0: $\mu = 83$ kg. H_1: $\mu < 83$ kg.
Test statistic: $t = -5.524$. Critical value: $t = -2.453$. P-value < 0.005 (Tech: 0.0000). Reject H_0. There is sufficient evidence to support the claim that male college students have a mean weight that is less than the 83 kg mean weight of males in the general population.

```
MINITAB
Test of mu = 83 vs < 83
Variable   N    Mean    StDev   SE Mean    T       P
WTSEP      32   72.72   10.53   1.86      -5.52    0.000
```

29. H_0: $\mu = 24$. H_1: $\mu < 24$. Test statistic: $z = \dfrac{19.6 - 24}{3.8/\sqrt{40}} = -7.32$. Critical value: $z = -1.645$.

P-value $= P(z < -7.32) = 0.0001$ (Tech: 0.0000). Reject H_0. There is sufficient evidence to support the claim that Chips Ahoy reduced-fat cookies have a mean number of chocolate chips that is less than 24 (but this does not provide conclusive evidence of reduced fat).

31. H_0: $\mu = 33$ years. H_1: $\mu \neq 33$ years. Test statistic: $z = \dfrac{35.9 - 33}{11.1/\sqrt{82}} = 2.37$. Critical values: $z = \pm 2.575$. P-value $= 2 \cdot P(z > 2.37) = 0.0178$ (Tech: 0.0180). Fail to reject H_0. There is not sufficient evidence to warrant rejection of the claim that the mean age of actresses when they win Oscars is 33 years.

33. $A = \dfrac{1.645(8 \cdot 149 + 3)}{8 \cdot 149 - 1} = 1.6505247$. The approximation yields a critical value of

$t = \sqrt{149\left(e^{1.6505247^2/149} - 1\right)} = 1.655$, which is the same as the result from STATDISK or a TI-83/84 Plus calculator.

35. a. The power of 0.4274 shows that there is a 42.74% chance of supporting the claim that $\mu < 1$ W/kg when the true mean is actually 0.80 W/kg. This value of power is not very high, and it shows that the hypothesis test is not very effective in recognizing that the mean is less than 1.00 W/kg when the actual mean is 0.80 W/kg.

    b. $\beta = 0.5726$. The probability of a type II error is 0.5726. That is, there is a 0.5726 probability of making the mistake of not supporting the claim that $\mu < 1$ W/kg when in reality the population mean is 0.80 W/kg.

Section 8-5

1. a. The mean waiting time remains the same.
   b. The variation among waiting times is lowered.
   c. Because customers all have waiting times that are roughly the same, they experience less stress and are generally more satisfied. Customer satisfaction is improved.
   d. The single line is better because it results in lower variation among waiting times, so a hypothesis test of a claim of a lower standard deviation is a good way to verify that the variation is lower with a single waiting line.

3. Use a 90% confidence interval. The conclusion based on the 90% confidence interval will be the same as the conclusion from a hypothesis test using the P-value method or the critical value method.

5. H_0: $\sigma = 0.15$ oz. H_1: $\sigma < 0.15$ oz. Test statistic: $\chi^2 = \dfrac{(36-1)0.11^2}{0.15^2} = 18.822$. Critical value of χ^2 is

between 18.493 and 26.509, so it is estimated to be 22.501 (Tech: 22.465). P-value < 0.05 (Tech: 0.0116). Reject H_0. There is sufficient evidence to support the claim that the population of volumes has a standard deviation less than 0.15 oz.

    MINITAB

Method	Chi-Square	DF	P-Value
Standard	18.82	35	0.012

7. H_0: $\sigma = 0.0230$ g. H_1: $\sigma < 0.0230$ g. Test statistic: $\chi^2 = \dfrac{(37-1)0.01648^2}{0.0230^2} = 18.483$. Critical value of χ^2

is between 18.493 and 26.509, so it is estimated to be 22.501 (Tech: 23.269). P-value < 0.05 (Tech: 0.0069). Reject H_0. There is sufficient evidence to support the claim that the population of weights has a standard deviation less than the specification of 0.0230 g.

    MINITAB

Method	Chi-Square	DF	P-Value
Standard	18.48	36	0.007

9. The data appear to be from a normally distributed population. H_0: $\sigma = 10$ bpm. H_1: $\sigma \ne 10$ bpm. Test

statistic: $\chi^2 = \dfrac{(40-1)10.3^2}{10^2} = 41.375$. Critical value of $\chi^2 = 24.433$ and $\chi^2 = 59.342$ (approximately).

P-value > 0.20 (Tech: 0.7347). Fail to reject H_0. There is not sufficient evidence to warrant rejection of the claim that pulse rates of men have a standard deviation equal to 10 beats per minute.

    MINITAB

Method	Chi-Square	DF	P-Value
Standard	41.38	39	0.735

11. H_0: $\sigma = 3.2$ mg. H_1: $\sigma \ne 3.2$ mg. Test statistic: $\chi^2 = \dfrac{(25-1)3.7^2}{3.2^2} = 32.086$. Critical values: $\chi^2 = 12.401$

and $\chi^2 = 39.364$. P-value > 0.20 (Tech: 0.2498). Fail to reject H_0. There is not sufficient evidence to support the claim that filtered 100-mm cigarettes have tar amounts with a standard deviation different from 3.2 mg. There is not enough evidence to conclude that filters have an effect.

    MINITAB

Method	Chi-Square	DF	P-Value
Standard	32.09	24	0.250

13. The data appear to be from a normally distributed population. H_0: $\sigma = 22.5$ years. H_1: $\sigma < 22.5$ years. Test

statistic: $\chi^2 = \dfrac{(15-1)7.67^2}{22.5^2} = 1.627$. Critical value: $\chi^2 = 4.660$. P-value < 0.005 (Tech: 0.0000). Reject

H_0. There is sufficient evidence to support the claim that the standard deviation of ages of all race car drivers is less than 22.5 years.

    MINITAB

Variable	Method	Chi-Square	DF	P-Value
Ages	Standard	1.63	14.00	0.000

15. The data appear to be from a normally distributed population. H_0: $\sigma = 32.2$ ft. H_1: $\sigma > 32.2$ ft. Test

statistic: $\chi^2 = \dfrac{(12-1)52.4^2}{32.2^2} = 29.176$. Critical value: $\chi^2 = 19.675$. P-value = 0.0021. Reject H_0. There is

sufficient evidence to support the claim that the new production method has errors with a standard deviation greater than 32.2 ft. The variation appears to be greater than in the past, so the new method appears to be worse, because there will be more altimeters that have larger errors. The company should take immediate action to reduce the variation.

MINITAB Variable	Method	Chi-Square	DF	P-Value
Errors	Standard	29.18	11.00	0.002

17. The data appear to be from a normally distributed population. H_0: $\sigma = 0.15$ oz. H_1: $\sigma < 0.15$ oz. Test

statistic: $\chi^2 = \dfrac{(36-1)0.0809^2}{0.15^2} = 10.173$. Critical value of χ^2 is between 18.493 and 26.509, so it is

estimated to be 22.501 (Tech: 22.465). P-value < 0.01 (Tech: 0.0000). Reject H_0. There is sufficient evidence to support the claim that the population of volumes has a standard deviation less than 0.15 oz.

MINITAB Variable	Method	Chi-Square	DF	P-Value
CKDTVOL	Standard	10.17	35.00	0.000

19. Critical $\chi^2 = \dfrac{1}{2}\left(-1.645 + \sqrt{2 \cdot 35 - 1}\right)^2 = 22.189$, which is reasonably close to the value of 22.465 obtained

from STATDISK and Minitab.

Chapter Quick Quiz

1. H_0: $\mu = 0$ sec. H_1: $\mu \neq 0$ sec.

2. a.   Two-tailed.

    b.   Student t.

3. a.   Fail to reject H_0.

    b.   There is not sufficient evidence to warrant rejection of the claim that the sample is from a population with a mean equal to 0 sec.

4. There is a loose requirement of a normally distributed population in the sense that the test works reasonably well if the departure from normality is not too extreme.

5. a. H_0: $p = 0.5$. H_1: $p > 0.5$.

    b. $z = \dfrac{0.64 - 0.5}{\sqrt{\frac{(0.5)(0.5)}{511}}} = 6.33$

    c.   P-value = 0.0000000001263996. There is sufficient evidence to support the claim that the majority of adults are in favor of the death penalty for a person convicted of murder.

6. $= 2 \cdot P(z < -2.00) = 0.0456$ (Tech: 0.0455)

7. The only true statement is the one given in part (a).

8. No. All critical values of x2 are greater than zero.

9. True.   10. False.

Review Exercises

1. a. False.

    b. True.

    c. False.

    d. False.

    e. False.

2. H_0: $p = \frac{2}{3}$. H_1: $p \neq \frac{2}{3}$. Test statistic: $z = \dfrac{\frac{657}{1010} - \frac{2}{3}}{\sqrt{\frac{\left(\frac{2}{3}\right)\left(\frac{1}{3}\right)}{1010}}} = -1.09$. Critical values: $z = \pm 2.575$ (Tech: ± 2.576).

    P-value $= 2 \cdot P(z < -1.09) = 0.2758$ (Tech: 0.2756). Fail to reject H_0. There is not sufficient evidence to warrant rejection of the claim that 2/3 of adults are satisfied with the amount of leisure time that they have.

    ```
    MINITAB
    Test of p = 0.666667 vs p not = 0.666667
    Sample    X      N  Sample p        95% CI        Z-Value  P-Value
    1        657   1010  0.650495  (0.621089, 0.679901)  -1.09    0.27
    ```

3. H_0: $p = 0.75$. H_1: $p > 0.75$. Test statistic: $z = \dfrac{\frac{678}{737} - 0.75}{\sqrt{\frac{(0.75)(0.25)}{737}}} = 10.65$ or $z = 10.66$ (if using $\hat{p} = 0.92$).

    Critical value: $z = \pm 2.33$. P-value = 0.0001 (Tech: 0.0000). Reject H_0. There is sufficient evidence to support the claim that more than 75% of us do not open unfamiliar e-mail and instant-message links. Given that the results are based on a voluntary response sample, the results are not necessarily valid.

    ```
    MINITAB
    Test of p = 0.75 vs p > 0.75
    Sample    X     N  Sample p  Z-Value  P-Value
    1        678   737  0.919946   10.65    0.000
    ```

4. H_0: $\mu = 3369$ g. H_1: $\mu < 3369$ g. Test statistic: $t = \dfrac{3245 - 567}{446/\sqrt{81}} = -19.962$. Critical value: $t = -2.328$

    (approximately). P-value < 0.005 (Tech: 0.0000). Reject H_0. There is sufficient evidence to support the claim that the mean birth weight of Chinese babies is less than the mean birth weight of 3369 g for Caucasian babies.

5. H_0: $\sigma = 567$ g. H_1: $\sigma \neq 567$ g. Test statistic: $\chi^2 = \dfrac{(81-1)466^2}{567^2} = 54.038$. Critical values of $\chi^2 = 51.172$

    and $\chi^2 = 116.321$. P-value is between 0.02 and 0.05 (Tech: 0.0229). Fail to reject H_0. There is not sufficient evidence to warrant rejection of the claim that the standard deviation of birth weights of Chinese babies is equal to 567 g.

    ```
    MINITAB
    Method    Chi-Square    DF    P-Value
    Standard     54.04       80     0.023
    ```

6. H_0: $\mu = 1.5$ mg/m^3. H_1: $\mu > 1.5$ mg/m^3. Test statistic: $t = 0.049$. Critical value: $t = 2.015$.

    P-value > 0.10 (Tech: 0.4814). Fail to reject H_0. There is not sufficient evidence to support the claim that the sample is from a population with a mean greater than the EPA standard of 1.5 mg/m^3. Because the sample value of 5.40 mg/m^3 appears to be an outlier and because a normal quantile plot suggests that the sample data are not from a normally distributed population, the requirements of the hypothesis test are not satisfied, and the results of the hypothesis test are therefore questionable.

6.   (continued)

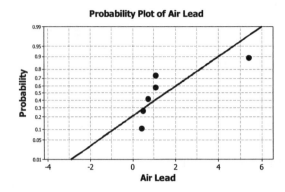

       MINITAB
       Test of mu = 1.5 vs > 1.5

Variable	N	Mean	StDev	SE Mean	T	P
Air Lead	6	1.538	1.914	0.781	0.05	0.481

7. H_0: $\mu = 25$. H_1: $\mu \neq 25$. Test statistic: $t = \dfrac{24.2 - 25}{14.1/\sqrt{100}} = -0.567$. Critical values: $t = \pm 1.984$

(approximately). P-value > 0.20 (Tech: 0.5717). Fail to reject H_0. There is not sufficient evidence to warrant rejection of the claim that the sample is selected from a population with a mean equal to 25.

       MINITAB
       Test of mu = 25 vs not = 25

N	Mean	StDev	SE Mean	95% CI	T	P
100	24.20	14.10	1.41	(21.40, 27.00)	-0.57	0.572

8.   a.   A type I error is the mistake of rejecting a null hypothesis when it is actually true. A type II error is the mistake of failing to reject a null hypothesis when in reality it is false.

    b.   Type I error: Reject the null hypothesis that the mean of the population is equal to 25 when in reality, the mean is actually equal to 25. Type II error: Fail to reject the null hypothesis that the population mean is equal to 25 when in reality, the mean is actually different from 25.

9.   The χ^2 test has a reasonably strict requirement that the sample data must be randomly selected from a population with a normal distribution, but the numbers are selected in such a way that they are all equally likely, so the population has a uniform distribution instead of the required normal distribution. Because the requirements are not all satisfied, the χ^2 2 test should not be used.

10.   The sample data meet the loose requirement of having a normal distribution. H_0: $\mu = 1000$ HIC. H_1: $\mu < 1000$ HIC. Test statistic: $t = -10.177$. Critical value: $t = -3.747$. P-value < 0.005 (Tech: 0.0003). Reject H_0. There is sufficient evidence to support the claim that the population mean is less than 1000 HIC. The results suggest that the population mean is less than 1000 HIC, so they appear to satisfy the specified requirement.

       MINITAB
       Test of mu = 1000 vs < 1000

Variable	N	Mean	StDev	SE Mean	T	P
Booster	5	653.8	76.1	34.0	-10.18	0.000

Cumulative Review Exercises

1.   a. $\bar{x} = 53.3$ words

    b.   Median = 52.0 words

    c. $s = 15.7$ words

    d. $s^2 = 245.1$ words2

    e.   Range = 45 words

2. a. Ratio.

    b. Discrete.

    c. The sample is a simple random sample if it was selected in such a way that all possible samples of the same size have the same chance of being selected.

3. $42.1 \text{ words} < \mu < 64.5 \text{ words}$

    MINITAB
    Variable N Mean StDev SE Mean   95% CI
    X   10 53.30 15.66   4.95   (42.10, 64.50)

4. H_0: $\mu = 48.0$ words. H_1: $\mu > 48.0$ words. Test statistic: $t = \dfrac{53.3 - 48.0}{15.7/\sqrt{10}} = 1.070$. Critical value: $t = 1.833$.

    P-value > 0.10 (Tech: 0.1561). Fail to reject H_0. There is not sufficient evidence to support the claim that the mean number of words on a page is greater than 48.0. There is not enough evidence to support the claim that there are more than 70,000 words in the dictionary.

    MINITAB
    Test of mu = 48 vs > 48
    N   Mean StDev SE Mean   T   P
    10   53.30 15.70   4.96   1.07 0.157

5. a. $z = \dfrac{38.8 - 36.0}{1.4} = 2$; $P(z > 2) = 2.28\%$.

    b. 98th percentile: $x = \mu + z \cdot \sigma = 36.0 + 2.054 \cdot 1.4 = 38.9$ in.

    c. $z = \dfrac{37.0 - 36.0}{1.4/\sqrt{4}} = 1.43$; $P(z < 1.43) = 92.36\%$. (Tech: 0.9234)

6. a. $(0.125)^3 = 0.00195$. It is unlikely because the probability of the event occurring is so small.

    b. $(0.097)(0.125) = 0.0121$ c. $1 - (0.875)^5 = 0.487$

7. No. The distribution is very skewed. A normal distribution would be approximately bell-shaped, but the displayed distribution is very far from being bell-shaped.

8. Because the vertical scale starts at 7000 and not at 0, the difference between the number of males and the number of females is exaggerated, so the graph is deceptive by creating the wrong impression that there are many more male graduates than female graduates.

9. a. $0.372(1003) = 373$

    b. $34.2\% < p < 40.2\%$

    MINITAB
    Sample   X   N Sample p   95% CI
    1   373 1003 0.371884   (0.341974, 0.401795)

    c. Yes. With test statistic $z = -8.11$ and with a P-value close to 0, there is sufficient evidence to support the claim that less than 50% of adults answer "yes."

    MINITAB
    Test of p = 0.5 vs p < 0.5
    Sample   X   N Sample p   Z-Value P-Value
    1   373 1003 0.371884   -8.11   0.000

    d. The required sample size depends on the confidence level and the sample proportion, not the population size.

10. H_0: $p = 0.5$. H_1: $p < 0.5$. Test statistic: $z = \dfrac{0.372 - 0.5}{\sqrt{\frac{(0.5)(0.5)}{1003}}} = -8.11$. Critical value: $z = -2.33$. P-value

$= P(z < -8.11) = 0.0001$ (Tech: 0.0000). Reject H_0. There is sufficient evidence to support the claim that fewer than 50% of Americans say that they have a gun in their home.

Chapter 9: Inferences from Two Samples

Section 9-2

1. The samples are simple random samples that are independent. For each of the two groups, the number of successes is at least 5 and the number of failures is at least 5. (Depending on what we call a success, the four numbers are 33, 115, 201,229 and 200,745 and all of those numbers are at least 5.) The requirements are satisfied.

3. a. $H_0: p_1 = p_2 . H_1: p_1 < p_2 .$

    b.   If the P-value is less than 0.001 we should reject the null hypothesis and conclude that there is sufficient evidence to support the claim that the rate of polio is less for children given the Salk vaccine than it is for children given a placebo.

5. Test statistic: $z = -12.39$ (rounded). The P-value of 3.137085E–35 is 0.0000 when rounded to four decimal places. There is sufficient evidence to warrant rejection of the claim that the vaccine has no effect.

For Exercises 7 – 17, assume that the data fit the requirements for the statistical methods for two proportions unless otherwise indicated.

7. a. $H_0: p_1 = p_2 . H_1: p_1 > p_2 .$ Test statistic: $z = 6.44$. Critical value: $z = 2.33$. P-value: 0.0001 (Tech: 0.0000). Reject H_0. There is sufficient evidence to support the claim that the proportion of people over 55 who dream in black and white is greater than the proportion for those under 25.

        MINITAB
        Difference = p (1) - p (2)
        Test for difference = 0 (vs > 0): Z = 6.44 P-Value = 0.000

    b.   98% CI: $0.117 < p_1 - p_2 < 0.240$. Because the confidence interval limits do not include 0, it appears that the two proportions are not equal. Because the confidence interval limits include only positive values, it appears that the proportion of people over 55 who dream in black and white is greater than the proportion for those under 25.

        MINITAB
        Difference = p (1) - p (2)
        98% CI for difference: (0.116836, 0.240360)

    c.   The results suggest that the proportion of people over 55 who dream in black and white is greater than the proportion for those under 25, but the results cannot be used to verify the cause of that difference.

9. a. $H_0: p_1 = p_2 . H_1: p_1 > p_2 .$ Test statistic: $z = 6.11$. Critical value: $z = 1.645$. P-value: 0.0001 (Tech: 0.0000). Reject H_0. There is sufficient evidence to support the claim that the fatality rate is higher for those not wearing seat belts.

        MINITAB
        Test for difference = 0 (vs > 0): Z = 6.11 P-Value = 0.000

    b.   90% CI: $0.00556 < p_1 - p_2 < 0.0122$. Because the confidence interval limits do not include 0, it appears that the two fatality rates are not equal. Because the confidence interval limits include only positive values, it appears that the fatality rate is higher for those not wearing seat belts.

        MINITAB
        Difference = p (1) - p (2)
        90% CI for difference: (0.00558525, 0.0122561)

    c.   The results suggest that the use of seat belts is associated with lower fatality rates than not using seat belts.

11. a. H_0: $p_1 = p_2$. H_1: $p_1 \neq p_2$. Test statistic: $z = 0.57$. Critical values: $z = \pm 1.96$. P-value: 0.5686 (Tech: 0.5720). Fail to reject H_0. There is not sufficient evidence to support the claim that echinacea treatment has an effect.

>   MINITAB
>   Difference = p (1) - p (2)
>   Test for difference = 0 (vs not = 0): Z = 0.57 P-Value = 0.572

   b.   95% CI: $-0.0798 < p_1 - p_2 < 0.149$. Because the confidence interval limits do contain 0, there is not a significant difference between the two proportions. There is not sufficient evidence to support the claim that echinacea treatment has an effect.

>   MINITAB
>   Difference = p (1) - p (2)
>   95% CI for difference: (-0.0798112, 0.148851)

   c.   Echinacea does not appear to have a significant effect on the infection rate. Because it does not appear to have an effect, it should not be recommended.

13. a. H_0: $p_1 = p_2$. H_1: $p_1 \neq p_2$. Test statistic: $z = 0.40$. Critical values: $z = \pm 1.96$. P-value: 0.6892 (Tech: 0.6859). Fail to reject H_0. There is not sufficient evidence to warrant rejection of the claim that men and women have equal success in challenging calls.

>   MINITAB
>   Difference = p (1) - p (2)
>   Test for difference = 0 (vs not = 0): Z = 0.40 P-Value = 0.686

   b.   95% CI: $-0.0318 < p_1 - p_2 < 0.0484$. Because the confidence interval limits contain 0, there is not a significant difference between the two proportions. There is not sufficient evidence to warrant rejection of the claim that men and women have equal success in challenging calls.

>   MINITAB
>   Difference = p (1) - p (2)
>   95% CI for difference: (-0.0318350, 0.0484421)

   c.   It appears that men and women have equal success in challenging calls.

15. a. H_0: $p_1 = p_2$. H_1: $p_1 > p_2$. Test statistic: $z = 9.97$. Critical value: $z = 2.33$. P-value: 0.0001 (Tech: 0.0000). Reject H_0. There is sufficient evidence to support the claim that the cure rate with oxygen treatment is higher than the cure rate for those given a placebo. It appears that the oxygen treatment is effective.

>   MINITAB
>   Difference = p (1) - p (2)
>   Test for difference = 0 (vs > 0): Z = 9.97 P-Value = 0.000

   b.   98% CI: $0.467 < p_1 - p_2 < 0.687$. Because the confidence interval limits do not include 0, it appears that the two cure rates are not equal. Because the confidence interval limits include only positive values, it appears that the cure rate with oxygen treatment is higher than the cure rate for those given a placebo. It appears that the oxygen treatment is effective.

>   MINITAB
>   Difference = p (1) - p (2)
>   98% CI for difference: (0.467454, 0.687321)

   c.   The results suggest that the oxygen treatment is effective in curing cluster headaches.

17. a. H_0: $p_1 = p_2$. H_1: $p_1 < p_2$. Test statistic: $z = -1.17$. Critical value: $z = -2.33$. P-value: 0.1210 (Tech: 0.1214). Fail to reject H_0. There is not sufficient evidence to support the claim that the rate of left-handedness among males is less than that among females.

>   MINITAB
>   Difference = p (1) - p (2)
>   Test for difference = 0 (vs < 0): Z = -1.17 P-Value = 0.121

17. (continued)

    b. 98% CI: $-0.0849 < p_1 - p_2 < 0.0265$ (Tech: $-0.0848 < p_1 - p_2 < 0.0264$). Because the confidence interval limits include 0, there does not appear to be a significant difference between the rate of left-handedness among males and the rate among females. There is not sufficient evidence to support the claim that the rate of left-handedness among males is less than that among females.

        MINITAB
        Difference = p (1) - p (2)
        98% CI for difference: (-0.0847744, 0.0264411)

    c. The rate of left-handedness among males does not appear to be less than the rate of left-handedness among females.

19. a. $0.0227 < p_1 - p_2 < 0.217$; because the confidence interval limits do not contain 0, it appears that $p_1 = p_2$ can be rejected.

        MINITAB
        Difference = p (1) - p (2)
        95% CI for difference: (0.0227099, 0.217290)

    b. $0.491 < p_1 < 0.629$; $0.371 < p_2 < 0.509$; because the confidence intervals do overlap, it appears that $p_1 = p_2$ cannot be rejected.

        MINITAB

Sample	X	N	Sample p	95% CI
1	112	200	0.560000	(0.488250, 0.629944)
2	88	200	0.440000	(0.370056, 0.511750)

    c. H_0: $p_1 = p_2$. H_1: $p_1 \neq p_2$. Test statistic: $z = 2.40$. P-value: 0.0164. Critical values: $z = \pm 1.96$. Reject H_0. There is sufficient evidence to reject $p_1 = p_2$.

        MINITAB
        Difference = p (1) - p (2)
        Test for difference = 0 (vs not = 0): Z = 2.40 P-Value = 0.016

    d. Reject $p_1 = p_2$. Least effective: Using the overlap between the individual confidence intervals.

21. $n = \dfrac{z_{\alpha/2}^2}{2E^2} = \dfrac{1.645^2}{2 \cdot 0.02^2} = 3383$ (Tech: 3382)

Section 9-3

1. Independent: b, d, e

3. Because the confidence interval does not contain 0, it appears that there is a significant difference between the mean height of women and the mean height of men. Based on the confidence interval, it appears that the mean height of men is greater than the mean height of women.

5. a. H_0: $\mu_1 = \mu_2$. H_1: $\mu_1 \neq \mu_2$. Test statistic: $t = -2.979$. Critical values: $t = \pm 2.032$ (Tech: ± 2.002). P-value < 0.01 (Tech: 0.0042). Reject H_0. There is sufficient evidence to warrant rejection of the claim that the samples are from populations with the same mean. Color does appear to have an effect on creativity scores. Blue appears to be associated with higher creativity scores.

        MINITAB
        Difference = mu (1) - mu (2)
        T-Test of difference = 0 (vs not =): T-Value = -2.98 P-Value = 0.004 DF = 58

    b. 95% CI: $-0.98 < \mu_1 - \mu_2 < -0.18$ (Tech: $-0.97 < \mu_1 - \mu_2 < -0.19$)

        MINITAB
        Difference = mu (1) - mu (2)
        95% CI for difference: (-0.970, -0.190)

7. a. H_0: $\mu_1 = \mu_2$. H_1: $\mu_1 > \mu_2$. Test statistic: $t = 0.132$. Critical value: $t = 1.729$. P-value > 0.10 (Tech: 0.4480). Fail to reject H_0. There is not sufficient evidence to support the claim that the magnets are effective in reducing pain. It is valid to argue that the magnets might appear to be effective if the sample sizes are larger.

    MINITAB
    Difference = mu (1) - mu (2)
    T-Test of difference = 0 (vs >): T-Value = 0.13 P-Value = 0.448 DF = 33

  b.   90% CI: $-0.61 < \mu_1 - \mu_2 < 0.71$ (Tech: $-0.59 < \mu_1 - \mu_2 < 0.69$)

    MINITAB
    Difference = mu (1) - mu (2)
    90% CI for difference: (-0.592, 0.692)

9. a.   The sample data meet the loose requirement of having a normal distribution. H_0: $\mu_1 = \mu_2$. H_1: $\mu_1 > \mu_2$. Test statistic: $t = 0.852$. Critical value: $t = 2.426$ (Tech: 2.676). P-value > 0.10 (Tech: 0.2054). Fail to reject H_0. There is not sufficient evidence to support the claim that men have a higher mean body temperature than women.

    MINITAB
    Difference = mu (1) - mu (2)
    T-Test of difference = 0 (vs >): T-Value = 0.85 P-Value = 0.206 DF = 12

  b.   98% CI: $-0.54\,°\mathrm{F} < \mu_1 - \mu_2 < 1.02\,°\mathrm{F}$ (Tech: $-0.51\,°\mathrm{F} < \mu_1 - \mu_2 < 0.99\,°\mathrm{F}$)

    MINITAB
    Difference = mu (1) - mu (2)
    98% CI for difference: (-0.515, 0.995)

11. a. H_0: $\mu_1 = \mu_2$. H_1: $\mu_1 < \mu_2$. Test statistic: $t = -3.547$. Critical value: $t = -2.462$ (Tech: –2.392). P-value < 0.005 (Tech: 0.0004). Reject H_0. There is sufficient evidence to support the claim that the mean maximal skull breadth in 4000 b.c. is less than the mean in a.d. 150.

    MINITAB
    Difference = mu (1) - mu (2)
    T-Test of difference = 0 (vs <): T-Value = -3.55 P-Value = 0.000 DF = 57

  b.   98% CI: $-8.13\text{ mm} < \mu_1 - \mu_2 < -1.47\text{ mm}$ (Tech: $-8.04\text{ mm} < \mu_1 - \mu_2 < -1.56\text{ mm}$)

    MINITAB
    Difference = mu (1) - mu (2)
    98% CI for difference: (-8.04, -1.56)

13. a. H_0: $\mu_1 = \mu_2$. H_1: $\mu_1 < \mu_2$. Test statistic: $t = -3.142$. Critical value: $t = -2.462$ (Tech: –2.403). P-value < 0.005 (Tech: 0.0014). Reject H_0. There is sufficient evidence to support the claim that students taking the nonproctored test get a higher mean than those taking the proctored test.

    MINITAB
    Difference = mu (1) - mu (2)
    T-Test of difference = 0 (vs <): T-Value = -3.17 P-Value = 0.001 DF = 49

  b.   98% CI: $-25.54 < \mu_1 - \mu_2 < -3.10$ (Tech: $-25.27 < \mu_1 - \mu_2 < -3.37$)

    MINITAB
    Difference = mu (1) - mu (2)
    98% CI for difference: (-25.27, -3.37)

15. a. H_0: $\mu_1 = \mu_2$. H_1: $\mu_1 \neq \mu_2$. Test statistic: $t = 1.274$. Critical values: $t = \pm 2.023$ (Tech: ± 1.994). P-value > 0.20 (Tech: 0.2066). Fail to reject H_0. There is not sufficient evidence to warrant rejection of the claim that males and females have the same mean BMI.

    MINITAB
    Difference = mu (1) - mu (2)
    T-Test of difference = 0 (vs not =): T-Value = 1.27 P-Value = 0.207 DF = 71

15. (continued)

    b. 95% CI: $-1.08 < \mu_1 - \mu_2 < 4.76$ (Tech: $-1.04 < \mu_1 - \mu_2 < 4.72$)

> MINITAB
> Difference = mu (1) - mu (2)
> 95% CI for difference: (-1.04, 4.72)

17. a. H_0: $\mu_1 = \mu_2$. H_1: $\mu_1 > \mu_2$. Test statistic: $t = 0.089$. Critical value: $t = 1.725$ (Tech: 2.029).

        P-value > 0.10 (Tech: 0.4648.) Fail to reject H_0. There is not sufficient evidence to support the claim that the mean IQ score of people with medium lead levels is higher than the mean IQ score of people with high lead levels.

> MINITAB
> Variable   N   Mean   StDev
> LOW LEAD   22   87.23   14.29
> Difference = mu (1) - mu (2)
> T-Test of difference = 0 (vs >): T-Value = 0.09 P-Value = 0.464 DF = 35

    b. 90% CI: $-5.9 < \mu_1 - \mu_2 < 6.6$ (Tech: $-5.8 < \mu_1 - \mu_2 < 6.4$)

> MINITAB
> Difference = mu (1) - mu (2)
> Estimate for difference: 0.33
> 90% CI for difference: (-5.80, 6.45)

19. a. H_0: $\mu_1 = \mu_2$. H_1: $\mu_1 < \mu_2$. Test statistic: $t = -1.810$. Critical value: $t = -2.650$ (Tech: -2.574). P-value > 0.025 (Tech: 0.0442). Fail to reject H_0. There is not sufficient evidence to support the claim that the mean longevity for popes is less than the mean for British monarchs after coronation.

> MINITAB
> Difference = mu (Popes) - mu (Kings and Queens)
> T-Test of difference = 0 (vs <): T-Value = -1.81 P-Value = 0.045 DF = 16

    b. 98% CI: -23.6 years $< \mu_1 - \mu_2 < 4.4$ years (Tech: -23.2 years $< \mu_1 - \mu_2 < 4.0$ years)

> MINITAB
> Difference = mu (Popes) - mu (Kings and Queens)
> 98% CI for difference: (-23.28, 4.10)

21. H_0: $\mu_1 = \mu_2$. H_1: $\mu_1 \neq \mu_2$. Test statistic: $t = 32.773$. Critical values: $t = \pm 2.023$ (Tech: ± 1.994). P-value < 0.01 (Tech: 0.0000). Reject H_0. There is sufficient evidence to warrant rejection of the claim that the two populations have equal means. The difference is highly significant, even though the samples are relatively small.

> MINITAB
> Difference = mu (Pre-1964 Quarters) - mu (Post-1964 Quarters)
> T-Test of difference = 0 (vs not =): T-Value = 32.77 P-Value = 0.000 DF = 70

23. 0.03795 lb $< \mu_1 - \mu_2 < 0.04254$ lb (Tech: 0.03786 lb $< \mu_1 - \mu_2 < 0.04263$ lb). Because the confidence interval does not include 0, there appears to be a significant difference between the two population means. It appears that the cola in cans of regular Pepsi weighs more than the cola in cans of Diet Pepsi, and that is probably due to the sugar in regular Pepsi that is not in Diet Pepsi.

> MINITAB
>   N   Mean   StDev
> PPREGWT   36 0.82410 0.00570
> PPDIETWT   36 0.78386 0.00436
> Difference = mu (PPREGWT) - mu (PPDIETWT)
> 95% CI for difference: (0.03786, 0.04263)

25. a.   The sample data meet the loose requirement of having a normal distribution. H_0: $\mu_1 = \mu_2$. H_1: $\mu_1 > \mu_2$.
    Test statistic: $t = 1.046$. Critical value: $t = 2.381$ (Tech: 2.382). P-value > 0.10 (Tech: 0.1496). Fail
    to reject H_0. There is not sufficient evidence to support the claim that men have a higher mean body
    temperature than women.

    > MINITAB
    > Difference = mu (1) - mu (2)
    > T-Test of difference = 0 (vs >): T-Value = 1.05 P-Value = 0.150 DF = 68
    > Both use Pooled StDev = 0.6986

    b.   98% CI: $-0.31°F < \mu_1 - \mu_2 < 0.79°F$. The test statistic became larger, the P-value became smaller,
    and the confidence interval became narrower, so pooling had the effect of attributing more significance
    to the results.

    > MINITAB
    > Difference = mu (1) - mu (2)
    > 98% CI for difference: (-0.307, 0.787)
    > Both use Pooled StDev = 0.6986

27. H_0: $\mu_1 = \mu_2$. H_1: $\mu_1 \neq \mu_2$. Test statistic: $t = 15.322$. Critical values: $t = \pm 2.080$. P-value < 0.01 (Tech:
    0.0000). Reject H_0. There is sufficient evidence to warrant rejection of the claim that the two populations
    have the same mean.

    $$t = \frac{(0.049 - 0.000) - (\mu_1 - \mu_2)}{\sqrt{\dfrac{s_p^2}{22} + \dfrac{s_p^2}{22}}}; \quad s_p^2 = \frac{(22-1)0.015^2 + (22-1)0^2}{(22-1) + (22-1)} = 0.000125$$

29. a. H_0: $\mu_1 = \mu_2$. H_1: $\mu_1 < \mu_2$. Test statistic: $t = -3.002$. Critical value based on 68.9927614 degrees of
    freedom: $t = -2.381$ (Tech: -2.382). P-value < 0.005 (Tech: 0.0019). Reject H_0. There is sufficient
    evidence to support the claim that students taking the nonproctored test get a higher mean than those
    taking the proctored test.

    b. $-25.68 < \mu_1 - \mu_2 < -2.96$ (Tech: $-25.69 < \mu_1 - \mu_2 < -2.95$)

Section 9-4

1.   Parts (c) and (e) are true.

3.   The test statistic will remain the same. The confidence interval limits will be expressed in the equivalent
    values of km/L.

5. H_0: $\mu_d = 0$ cm. H_1: $\mu_d > 0$ cm. Test statistic: $t = 0.036$ (rounded). Critical value: $t = 1.692$.
    P-value > 0.10 (Tech: 0.4859). Fail to reject H_0. There is not sufficient evidence to support the claim that
    for the population of heights of presidents and their main opponents; the differences have a mean greater
    than 0 cm (with presidents tending to be taller than their opponents).

7.   a. $\bar{d} = -11.6$ years   b. $s_d = 17.2$ years

    c.   Test statistic $t = \dfrac{\bar{d} - \mu_d}{s_d / \sqrt{n}} = \dfrac{-11.6 - 0}{17.2 / \sqrt{5}} = -1.508$

    d. H_0: $\mu_d = 0$. H_1: $\mu_d \neq 0$. Critical values: $t = \pm 2.776$.

9. H_0: $\mu_d = 0$. H_1: $\mu_d \neq 0$. Test statistic: $t = \dfrac{-11.6 - 0}{17.21 / \sqrt{5}} = -1.507$. Critical values: $t = \pm 2.776$.
    P-value > 0.20 (Tech: 0.2063). Fail to reject H_0. There is not sufficient evidence to support the claim that
    there is a difference between the ages of actresses and actors when they win Oscars.

    > MINITAB
    > Paired T for Actress - Actor
    > T-Test of mean difference = 0 (vs not = 0): T-Value = -1.51 P-Value = 0.206

11. 1.0 min $< \mu_d < 12.0$ min. Because the confidence interval includes only positive values and does not include 0 min, it appears that the taxi-out times are greater than the corresponding taxi-in times, so there is sufficient evidence to support the claim of the flight operations manager that for flight delays, more of the blame is attributable to taxi-out times at JFK than taxi-in times at LAX.

> MINITAB
> Paired T for Out - In
> 90% CI for mean difference: (0.99, 12.01)

$\bar{d} = 6.5$ min; df $= 12 - 1 = 11$

$E = t_{\alpha/2} \cdot \dfrac{s_d}{\sqrt{n}} = 1.796 \cdot \dfrac{10.63}{\sqrt{12}} = 5.5$ min

13. H_0: $\mu_d = 0$. H_1: $\mu_d > 0$. Test statistic: $t = \dfrac{7279 - 0}{6913/\sqrt{6}} = 2.579$. Critical value: $t = 2.015$.

P-value < 0.025 (Tech: 0.0247). Reject H_0. There is sufficient evidence to support the claim that among couples, males speak more words in a day than females.

> MINITAB
> Paired T for Male - Female
> T-Test of mean difference = 0 (vs > 0): T-Value = 2.58 P-Value = 0.025

15. $-6.5 < \mu_d < -0.2$. Because the confidence interval does not include 0, it appears that there is sufficient evidence to warrant rejection of the claim that when the 13th day of a month falls on a Friday, the numbers of hospital admissions from motor vehicle crashes are not affected. Hospital admissions do appear to be affected.

> MINITAB
> Paired T for Friday the 6th - Friday the 13th
> 95% CI for mean difference: (-6.49, -0.17)

$\bar{d} = -3.33$ cm^3; df $= 6 - 1 = 5$

$E = t_{\alpha/2} \cdot \dfrac{s_d}{\sqrt{n}} = 2.571 \cdot \dfrac{3.01}{\sqrt{6}} = 3.2$ cm^3

17. H_0: $\mu_d = 0$. H_1: $\mu_d < 0$. Test statistic: $t = \dfrac{-1.57 - 0}{4.60/\sqrt{10}} = -1.080$. Critical value: $t = -1.833$.

P-value > 0.10 (Tech: 0.1540). Fail to reject H_0. There is not sufficient evidence to support the claim that *Harry Potter and the Half-Blood Prince* did better at the box office. After a few years, the gross amounts from both movies can be identified, and the conclusion can then be judged objectively without using a hypothesis test.

> MINITAB
> Paired T for Phoenix - Prince
> T-Test of mean difference = 0 (vs < 0): T-Value = -1.08 P-Value = 0.154

19. $0.69 < \mu_d < 5.56$. Because the confidence interval limits do not contain 0 and they consist of positive values only, it appears that the "before" measurements are greater than the "after" measurements, so hypnotism does appear to be effective in reducing pain.

> MINITAB
> Paired T for Before - After
> 95% CI for mean difference: (0.69, 5.56)

$\bar{d} = 3.13$; df $= 8 - 1 = 7$

$E = t_{\alpha/2} \cdot \dfrac{s_d}{\sqrt{n}} = 2.365 \cdot \dfrac{2.91}{\sqrt{8}} = 2.43$

21. H_0: $\mu_d = 0$. H_1: $\mu_d \neq 0$. Test statistic: $t = -5.553$. Critical values: $t = \pm 1.990$. P-value < 0.01 (Tech: 0.0000). Reject H_0. There is sufficient evidence to support the claim that there is a difference between the ages of actresses and actors when they win Oscars.

> MINITAB
> Paired T for Actresses - Actors
> T-Test of mean difference = 0 (vs not = 0): T-Value = -5.55 P-Value = 0.000

23. H_0: $\mu_d = 0$. H_1: $\mu_d < 0$. Test statistic: $t = -1.560$. Critical value of t is between -1.671 and -1.676 (Tech: -1.673). P-value > 0.05 (Tech: 0.0622). Fail to reject H_0. There is not sufficient evidence to support the claim that among couples, males speak fewer words in a day than females.

    MINITAB
    Paired T for M1 - F1
    95% upper bound for mean difference: 135
    T-Test of mean difference = 0 (vs < 0): T-Value = -1.56 P-Value = 0.062

25. H_0: $\mu_d = 6.8\,\text{kg}$. H_1: $\mu_d \neq 6.8\,\text{kg}$. Test statistic: $t = -11.833$. Critical values: $t = \pm 1.994$ (Tech: ± 1.997). P-value < 0.01 (Tech: 0.0000). Reject H_0. There is sufficient evidence to warrant rejection of the claim that $\mu_d = 6.8\,\text{kg}$. It appears that the "Freshman 15" is a myth, and college freshman might gain some weight, but they do not gain as much as 15 pounds.

    MINITAB
    Paired T for WTAPR - WTSEP
    T-Test of mean difference = 6.8 (vs not = 6.8): T-Value = -11.83 P-Value = 0.000

Section 9-5

1. a. No.

    b. No.

    c. The two samples have the same standard deviation (or variance).

3. The F test is very sensitive to departures from normality, which means that it works poorly by leading to wrong conclusions when either or both of the populations has a distribution that is not normal. The F test is not robust against sampling methods that do not produce simple random samples. For example, conclusions based on voluntary response samples could easily be wrong.

5. H_0: $\sigma_1 = \sigma_2$. H_1: $\sigma_1 \neq \sigma_2$. Test statistic: $F = 1.7341$. Upper critical F value is between 1.8752 and 2.0739 (Tech: 1.9611). P-value: 0.1081. Fail to reject H_0. There is not sufficient evidence to support the claim that weights of regular Coke and weights of regular Pepsi have different standard deviations.

7. H_0: $\sigma_1 = \sigma_2$. H_1: $\sigma_1 \neq \sigma_2$. Test statistic: $F = \dfrac{5.90^2}{5.48^2} = 1.1592$. Upper critical F value is between 1.8752 and 2.0739 (Tech: 1.9678). P-value: 0.6656. Fail to reject H_0. There is not sufficient evidence to warrant rejection of the claim that the samples are from populations with the same standard deviation. The background color does not appear to have an effect on the variation of word recall scores.

    MINITAB
    Test for Equal Variances
    F-Test (Normal Distribution)
    Test statistic = 1.16,   p-value = 0.666

9. H_0: $\sigma_1 = \sigma_2$. H_1: $\sigma_1 > \sigma_2$. Test statistic: $F = \dfrac{2.2^2}{0.72^2} = 9.3364$. Critical F value is between 12.0540 and 2.0960 (Tech: 2.0842). P-value: 0.0000. Reject H_0. There is sufficient evidence to support the claim that the treatment group has errors that vary more than the errors of the placebo group.

    MINITAB
    F-Test (Normal Distribution)
    Test statistic = 9.34,   p-value = 0.000

11. H_0: $\sigma_1 = \sigma_2$. H_1: $\sigma_1 > \sigma_2$. Test statistic: $F = \dfrac{1.4^2}{0.96^2} = 2.1267$. Critical F value is between 2.1555 and 2.2341 (Tech: 2.1682). P-value: 0.0543. Fail to reject H_0. There is not sufficient evidence to support the claim that those given a sham treatment (similar to a placebo) have pain reductions that vary more than the pain reductions for those treated with magnets.

> MINITAB
> Test for Equal Variances
> F-Test (Normal Distribution)
> Test statistic = 2.13,   p-value = 0.109

13. H_0: $\sigma_1 = \sigma_2$. H_1: $\sigma_1 > \sigma_2$. Test statistic: $F = \dfrac{10.6383^2}{5.2129^2} = 4.1648$. Critical F value is between 2.7876 and 2.8536 (Tech: 2.8179). P-value: 0.0130. Reject H_0. There is sufficient evidence to support the claim that amounts of strontium-90 from Pennsylvania residents vary more than amounts from New York residents.

> MINITAB
> Test for Equal Variances
> F-Test (Normal Distribution)
> Test statistic = 4.16,   p-value = 0.026

15. H_0: $\sigma_1 = \sigma_2$. H_1: $\sigma_1 \neq \sigma_2$. Test statistic: $F = \dfrac{6.06465^2}{6.04264^2} = 1.0073$. Upper critical F value: 4.0260. P-value: 0.9915. Fail to reject H_0. There is not sufficient evidence to warrant rejection of the claim that females and males have heights with the same amount of variation.

> MINITAB
> Test for Equal Variances
> F-Test (Normal Distribution)
> Test statistic = 0.99,   p-value = 0.992

17. H_0: $\sigma_1 = \sigma_2$. H_1: $\sigma_1 > \sigma_2$. Test statistic: $F = \dfrac{20.6883^2}{18.5281^2} = 1.2397$. Critical F value is between 1.6928 and 1.8409 (Tech: 1.7045). P-value: 0.2527. Fail to reject H_0. There is not sufficient evidence to support the claim that males have weights with more variation than females.

> MINITAB
> Test for Equal Variances
> F-Test (Normal Distribution)
> Test statistic = 1.25,   p-value = 0.494

19. a. No solution provided.

　　b. $c_1 = 4$, $c_2 = 0$

　　c. Critical value $= \dfrac{\log(0.05/2)}{\log\left(\dfrac{40}{40+40}\right)} = 5$.

　　d. Fail to reject $\sigma_1^2 = \sigma_2^2$.

21. $F_L = 0.2727$, $F_R = 2.8365$

Chapter Quick Quiz

1. H_0: $p_1 = p_2$. H_1: $p_1 \neq p_2$.

2. $\overline{p} = \dfrac{347 + 305}{386 + 359} = 0.875$

3. $2 \cdot P(z > 2.04) = 0.0414\ 0.0414$

4. $0.00172 < p_1 - p_2 < 0.0970$

5. Because the data consist of matched pairs, they are dependent.

6. H_0: $\mu_d = 0$. H_1: $\mu_d > 0$.

7. There is not sufficient evidence to support the claim that front repair costs are greater than the corresponding rear repair costs.

9. False.

10. True.

8. *F* distribution

Review Exercises

1. H_0: $p_1 = p_2$. H_1: $p_1 > p_2$. Test statistic: $z = 3.12$. Critical value: $z = 2.33$. P-value: 0.0009. Reject H_0. There is sufficient evidence to support a claim that the proportion of successes with surgery is greater than the proportion of successes with splinting. When treating carpal tunnel syndrome, surgery should generally be recommended instead of splinting.

    MINITAB
    Difference = p (1) - p (2)
    Test for difference = 0 (vs > 0): Z = 3.12 P-Value = 0.001

2. 98% CI: $0.0581 < p_1 - p_2 < 0.332$ (Tech: $0.0583 < p_1 - p_2 < 0.331$). The confidence interval limits do not contain 0; the interval consists of positive values only. This suggests that the success rate with surgery is greater than the success rate with splints.

    MINITAB
    Difference = p (1) - p (2)
    98% CI for difference: (0.0583369, 0.331496)

3. H_0: $p_1 = p_2$. H_1: $p_1 < p_2$. Test statistic: $z = -1.91$. Critical value: $z = -1.645$. P-value: 0.0281 (Tech: 0.0280). Reject H_0. There is sufficient evidence to support the claim that the fatality rate of occupants is lower for those in cars equipped with airbags.

    MINITAB
    Difference = p (1) - p (2)
    Test for difference = 0 (vs < 0): Z = -1.91 P-Value = 0.028

4. H_0: $\mu_d = 0$. H_1: $\mu_d > 0$. Test statistic: $t = 4.712$. Critical value: $t = 3.143$. P-value < 0.005 (Tech: 0.0016). Reject H_0. There is sufficient evidence to support the claim that flights scheduled 1 day in advance cost more than flights scheduled 30 days in advance. Save money by scheduling flights 30 days in advance.

    MINITAB
    Paired T for Flight scheduled one day in adv - Flight scheduled 30 days in adv
    T-Test of mean difference = 0 (vs > 0): T-Value = 4.71 P-Value = 0.002

5. H_0: $\mu_d = 0$. H_1: $\mu_d \neq 0$. Test statistic: $t = -0.574$. Critical values: $t = \pm 2.365$. P-value > 0.20 (Tech: 0.5840). Fail to reject H_0. There is not sufficient evidence to support the claim that there is a difference between self-reported heights and measured heights of females aged 12–16.

    MINITAB
    Paired T for Reported Height - Measured Height
    T-Test of mean difference = 0 (vs not = 0): T-Value = -0.57 P-Value = 0.584

6. H_0: $\mu_1 = \mu_2$. H_1: $\mu_1 > \mu_2$. Test statistic: $t = 2.879$. Critical value: $t = 2.429$ (Tech: 2.376). P-value < 0.005 (Tech: 0.0026). Reject H_0. There is sufficient evidence to support the claim that "stress decreases the amount recalled."

    MINITAB
    Difference = mu (1) - mu (2)
    T-Test of difference = 0 (vs >): T-Value = 2.88 P-Value = 0.003 DF = 76

7. 98% CI: $1.3 < \mu_1 - \mu_2 < 14.7$ (Tech: $1.4 < \mu_1 - \mu_2 < 14.6$). The confidence interval limits do not contain 0; the interval consists of positive values only. This suggests that the numbers of details recalled are lower for those in the stress population.

    MINITAB
    Difference = mu (1) - mu (2)
    98% CI for difference: (1.40, 14.60)

8. H_0: $p_1 = p_2$. H_1: $p_1 \neq p_2$. Test statistic: $z = -4.20$. Critical values: $z = \pm 2.575$. P-value: 0.0002 (Tech: 0.0000). Reject H_0. There is sufficient evidence to warrant rejection of the claim that the acceptance rate is the same with or without blinding. Without blinding, reviewers know the names and institutions of the abstract authors, and they might be influenced by that knowledge.

    MINITAB
    Difference = p (1) - p (2)
    Test for difference = 0 (vs not = 0): Z = -4.20 P-Value = 0.000

9. H_0: $\mu_1 = \mu_2$. H_1: $\mu_1 \neq \mu_2$. Test statistic: $t = 0.679$. Critical values: $t = \pm 2.014$ approximately (Tech: ± 1.985). P-value > 0.20 (Tech: 0.4988). Fail to reject H_0. There is not sufficient evidence to warrant rejection of the claim of no difference between the mean LDL cholesterol levels of subjects treated with raw garlic and subjects given placebos. Both groups appear to be about the same.

    MINITAB
    Difference = mu (1) - mu (2)
    T-Test of difference = 0 (vs not =): T-Value = 0.68 P-Value = 0.499 DF = 94

10. H_0: $\sigma_1 = \sigma_2$. H_1: $\sigma_1 \neq \sigma_2$. Test statistic: $F = 1.1480$. Upper critical F value is between 1.6668 and 1.8752 (Tech: 1.7799). P-value: 0.6372. Fail to reject H_0. There is not sufficient evidence to warrant rejection of the claim that the two populations have LDL levels with the same standard deviation.

    MINITAB
    F-Test (Normal Distribution)
    Test statistic = 1.15,   p-value = 0.637

Cumulative Review Exercises

1. a. Because the sample data are matched with each column consisting of heights from the same family, the data are dependent.

    b. Mean: 63.81 in.; median: 63.70 in.; mode: 62.2 in.; range: 8.80 in.; standard deviation: 2.73 in.; variance: 7.43 in^2

    c. Ratio

2. There does not appear to be a correlation or association between the heights of mothers and the heights of their daughters.

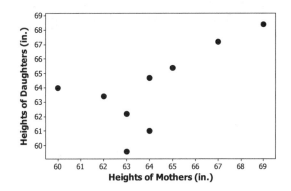

3. 61.86 in. $< \mu < 65.76$ in. We have 95% confidence that the limits of 61.86 in. and 65.76 in. actually contain the true value of the mean height of all adult daughters.

    MINITAB
    Variable   N   Mean   StDev SE Mean   95% CI
    Daughters 10   63.810   2.726   0.862   (61.860, 65.760)

4. H_0: $\mu_d = 0$. H_1: $\mu_d \neq 0$. Test statistic: $t = 0.283$. Critical values: $t = \pm 2.262$. P-value > 0.20 (Tech: 0.7834). Fail to reject H_0. There is not sufficient evidence to warrant rejection of the claim of no significant difference between the heights of mothers and the heights of their daughters.

    MINITAB
    Paired T for Heights of Mothers (in.) - Heights of Daughters (in.)
    T-Test of mean difference = 0 (vs not = 0): T-Value = 0.28 P-Value = 0.783

5. Because the points lie reasonably close to a straight-line pattern and there is no other pattern that is not a straight-line pattern and there are no outliers, the sample data appear to be from a population with a normal distribution.

6. $0.109 < p_1 < 0.150$. Because the entire range of values in the confidence interval lies below 0.20, the results do justify the statement that "fewer than 20% of Americans choose their computer and/or Internet access when identifying what they miss most when electrical power is lost."

    MINITAB
    Sample X   N   Sample p   95% CI
    1   134 1032   0.129845   (0.109337, 0.150353)

7. No. Because the Internet users chose to respond, we have a voluntary response sample, so the results are not necessarily valid.

8. $n = \dfrac{[z_{\alpha/2}]^2 \hat{p}\hat{q}}{E^2} = \dfrac{[2.17]^2 (0.25)}{0.02^2} = 2944$. The survey should not be conducted using only local phone numbers. Such a convenience sample could easily lead to results that are dramatically different from results that would be obtained by randomly selecting respondents from the entire population, not just those having local phone numbers.

9. a. $z = \dfrac{152.1 - 162.0}{6.6} = -1.5$; $P(z > -1.5) = 0.9332$.

    b. $z = \dfrac{152.1 - 162.0}{6.6/\sqrt{4}} = -3$; $P(z > -1.5) = 0.0.9987$.

    c.   80th percentile: $x = \mu + z \cdot \sigma = 162.0 + 0.842 \cdot 6.6 = 167.6$ cm.

10. No. Because the states have different population sizes, the mean cannot be found by adding the 50 state means and dividing the total by 50. The mean income for the U.S. population can be found by using a weighted mean that incorporates the population size of each state.

Chapter 10: Correlation and Regression

Section 10-2

1. r represents the value of the linear correlation computed by using the paired sample data. ρ represents the value of the linear correlation coefficient that would be computed by using all of the paired data in the population. The value of r is estimated to be 0 (because there is no correlation between sunspot numbers and the Dow Jones Industrial Average).

3. The headline is not justified because it states that increased salt consumption is the cause of higher blood pressure levels, but the presence of a correlation between two variables does not necessarily imply that one is the cause of the other. Correlation does not imply causality. A correct headline would be this: "Study Shows That Increased Salt Consumption Is Associated with Higher Blood Pressure."

5. H_0: $\rho = 0$. H_1: $\rho \neq 0$; Yes. With $r = 0.687$ and critical values of ± 0.312, there is sufficient evidence to support the claim that there is a linear correlation between the durations of eruptions and the time intervals to the next eruptions.

7. H_0: $\rho = 0$. H_1: $\rho \neq 0$; No. With $r = 0.149$ and a P-value of 0.681 (or critical values of ± 0.632), there is not sufficient evidence to support the claim that there is a linear correlation between the heights of fathers and the heights of their sons.

9. a.

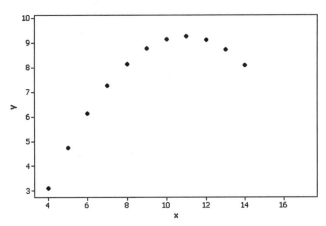

b. H_0: $\rho = 0$. H_1: $\rho \neq 0$; $r = 0.816$. Critical values: $r = \pm 0.602$. P-value = 0.002. There is sufficient evidence to support the claim of a linear correlation between the two variables.

   MINITAB
   Pearson correlation of x and y = 0.816
   P-Value = 0.002

c. The scatterplot reveals a distinct pattern that is not a straight line pattern.

11. a. There appears to be a linear correlation.

   b. H_0: $\rho = 0$. H_1: $\rho \neq 0$; $r = 0.906$. Critical values: $r = \pm 0.632$ (for a 0.05 significance level). There is a linear correlation.

   MINITAB
   Pearson correlation of x and y = 0.906
   P-Value = 0.000

   c. H_0: $\rho = 0$. H_1: $\rho \neq 0$; $r = 0$. Critical values: $r = \pm 0.666$ (for a 0.05 significance level). There does not appear to be a linear correlation.

   MINITAB
   Pearson correlation of x and y = 0.000
   P-Value = 1.000

   d. The effect from a single pair of values can be very substantial, and it can change the conclusion.

13. H_0: $\rho = 0$. H_1: $\rho \neq 0$; $r = -0.959$. Critical values: $r = \pm 0.878$. P-value = 0.010. There is sufficient evidence to support the claim that there is a linear correlation between weights of lemon imports from Mexico and U.S. car fatality rates. The results do not suggest any cause-effect relationship between the two variables.

> MINITAB
> Pearson correlation of Lemon Imports and
>   Fatality Rate = -0.959
> P-Value = 0.010

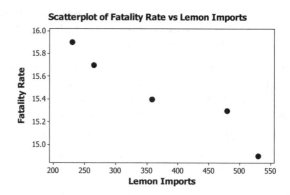

15. H_0: $\rho = 0$. H_1: $\rho \neq 0$; $r = 0.561$. Critical values: $r = \pm 0.632$. P-value = 0.091. There is not sufficient evidence to support the claim that there is a linear correlation between enrollment and burglaries. The results do not change if the actual enrollments are listed as 32,000, 31,000, 53,000, etc.

> MINITAB
> Pearson correlation of Enrollment and
>   Burglaries = 0.561
> P-Value = 0.091

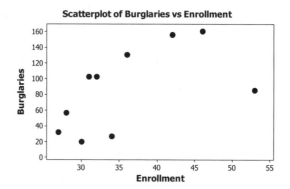

17. H_0: $\rho = 0$. H_1: $\rho \neq 0$; $r = 0.864$. Critical values: $r = \pm 0.666$. P-value = 0.003. There is sufficient evidence to support the claim that there is a linear correlation between court incomes and justice salaries. The correlation does not imply that court incomes directly affect justice salaries, but it does appear that justices might profit by levying larger fines, or perhaps justices with higher salaries impose larger fines.

> MINITAB
> Pearson correlation of Income and Salary
>   = 0.864
> P-Value = 0.003

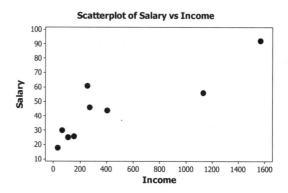

19. H_0: $\rho = 0$. H_1: $\rho \neq 0$; $r = 1.000$. Critical values: $r = \pm 0.811$. P-value = 0.000. There is sufficient evidence to support the claim that there is a linear correlation between amounts of redshift and distances to clusters of galaxies. Because the linear correlation coefficient is 1.000, it appears that the distances can be directly computed from the amounts of redshift.

MINITAB
Pearson correlation of Red and Dist
  = 1.000
P-Value = 0.000

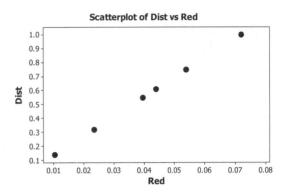

21. H_0: $\rho = 0$. H_1: $\rho \neq 0$; $r = 0.948$. Critical values: $r = \pm 0.811$. P-value = 0.004. There is sufficient evidence to support the claim of a linear correlation between the overhead width of a seal in a photograph and the weight of a seal.

MINITAB
Pearson correlation of Width and Weight
  = 0.948
P-Value = 0.004

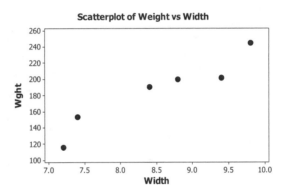

23. H_0: $\rho = 0$. H_1: $\rho \neq 0$; $r = 0.867$. Critical values: $r = \pm 0.878$. P-value = 0.057. There is not sufficient evidence to support the claim of a linear correlation between the systolic blood pressure measurements of the right and left arm.

MINITAB
Pearson correlation of Right Arm and Left Arm = 0.867
P-Value = 0.057

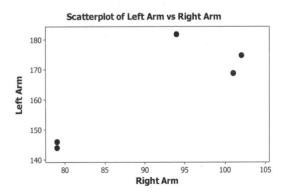

25. H_0: $\rho = 0$. H_1: $\rho \neq 0$; $r = 0.197$. Critical values: $r = \pm 0.707$. P-value = 0.640. There is not sufficient evidence to support the claim that there is a linear correlation between prices of regular gas and prices of premium gas. Because there does not appear to be a linear correlation between prices of regular and premium gas, knowing the price of regular gas is not very helpful in getting a good sense for the price of premium gas.

MINITAB
Pearson correlation of Reg and Prem
    = 0.197
P-Value = 0.640

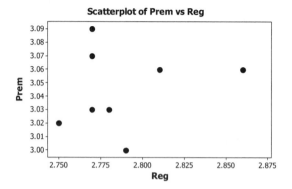

27. H_0: $\rho = 0$. H_1: $\rho \neq 0$; $r = 1.000$. Critical values: $r = \pm 0.707$. P-value = 0.000. There is sufficient evidence to support the claim that there is a linear correlation between diameters and circumferences. A scatterplot confirms that there is a linear association between diameters and volumes.

MINITAB
Pearson correlation of Diam and Circ
    = 1.000
P-Value = 0.000

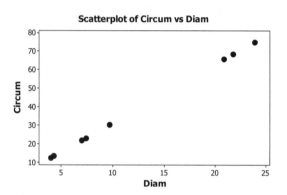

29. H_0: $\rho = 0$. H_1: $\rho \neq 0$; $r = -0.063$. Critical values: $r = \pm 0.444$. P-value = 0.791. There is not sufficient evidence to support the claim of a linear correlation between IQ and brain volume.

MINITAB
Pearson correlation of IQ and VOL = -0.063
P-Value = 0.791

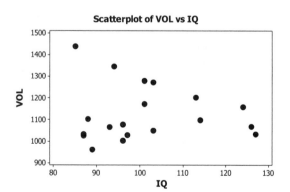

31. H_0: $\rho = 0$. H_1: $\rho \neq 0$; $r = 0.319$. Critical values: $r = \pm 0.254$ (approximately) (Tech: ± 0.263). P-value = 0.017. There is sufficient evidence to support the claim of a linear correlation between the numbers of words spoken by men and women who are in couple relationships.

MINITAB
Pearson correlation of M1 and F1 = 0.319
P-Value = 0.017

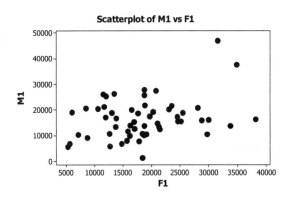

33. a. $r = 0.911$

MINITAB
Pearson correlation of y and x = 0.911
P-Value = 0.031

b. $r = 0.787$

MINITAB
Pearson correlation of y and x^2 = 0.787
P-Value = 0.114

c. $r = 0.9999$ (largest)

MINITAB
Pearson correlation of y and LOG(x)
    = 1.000
P-Value = 0.000

d. $r = 0.976$

MINITAB
Pearson correlation of y and SQRT(x)
    = 0.976
P-Value = 0.005

e. $r = -0.948$

MINITAB
Pearson correlation of y and 1/x
    = -0.948
P-Value = 0.014

Section 10-3

1. The symbol $\hat{y}$ represents the predicted pulse rate. The predictor variable represents height. The response variable represents pulse rate.

3. If r is positive, the regression line has a positive slope and rises from left to right. If r is negative, the slope of the regression line is negative and it falls from left to right.

5. The regression line fits the points well, so the best predicted time for an interval after the eruption is
$\hat{y} = 47.4 + 0.180(120) = 69$ min.

7. The regression line does not fit the points well, so the best predicted height is $\bar{y} = 68.0$ in.

9. $\hat{y} = 3.00 + 0.500x$. The data have a pattern that is not a straight line.

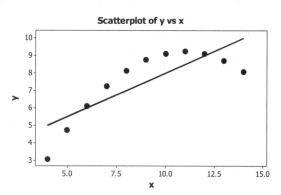

```
MINITAB
Predictor   Coef      SE Coef    T      P
Constant    3.001     1.125      2.67   0.026
x           0.5000    0.1180     4.24   0.002
```

11. a. $\hat{y} = 0.264 + 0.906x$

```
MINITAB
Predictor   Coef      SE Coef    T      P
Constant    0.2642    0.5649     0.47   0.653
x           0.9057    0.1499     6.04   0.000
```

b. $\hat{y} = 2 + 0x$ (or $\hat{y} = 2$)

```
MINITAB
Predictor   Coef      SE Coef    T      P
Constant    2.0000    0.8165     2.45   0.044
x          -0.0000    0.3780    -0.00   1.000
```

c.   The results are very different, indicating that one point can dramatically affect the regression equation.

13. $\hat{y} = 16.5 - 0.00282x$; The regression line fits the points well, so the best predicted value is

$\hat{y} = 16.5 - 0.00282(500) = 15.1$ fatalities per 100,000 population.

```
MINITAB
Predictor   Coef       SE Coef     T       P
Constant    16.4909    0.1880      87.70   0.000
Lemon      -0.00282    0.0004815  -5.86    0.010
```

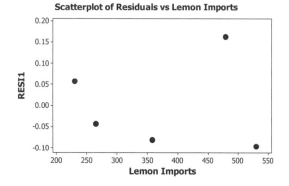

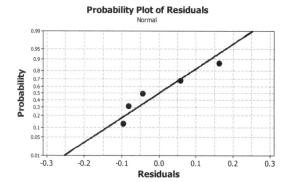

15. $\hat{y} = -36.8 + 3.47x$; The regression line does not fit the points well, so the best predicted value is $\bar{y} = 87.7$ burglaries. The predicted value is not close to the actual value of 329 burglaries.

MINITAB

Predictor	Coef	SE Coef	T	P
Constant	-36.77	66.50	-0.55	0.595
Enrollment	3.467	1.807	1.92	0.091

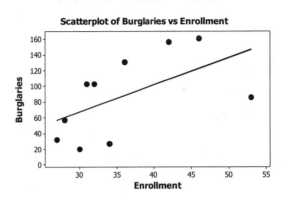

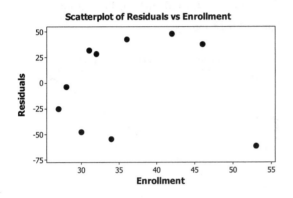

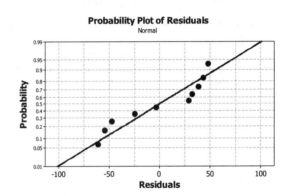

17. $\hat{y} = 27.7 + 0.0373x$; The best predicted value is $\hat{y} = 27.7 + 0.0373(83.941) = 30.8$, which represents $30,800. The predicted value is not very close to the actual salary of $26,088. The possible outliers might explain the inaccuracy.

MINITAB

Predictor	Coef	SE Coef	T	P
Constant	27.701	5.519	5.02	0.002
Income	0.03728	0.008201	4.55	0.003

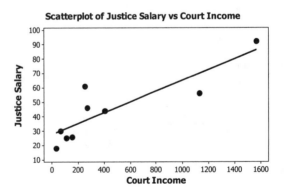

17. (continued)

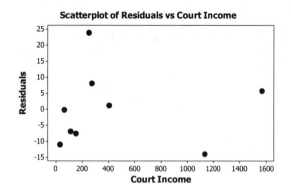

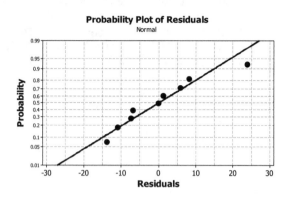

19. $\hat{y} = -0.00440 + 14.0x$; The best predicted value is $\hat{y} = -0.00440 + 14.0(0.0126) = 0.172$ billion light-years. The predicted value is very close to the actual distance of 0.18 light-years.

MINITAB

Predictor	Coef	SE Coef	T	P
Constant	-0.004396	0.00125	-3.51	0.025
Redshift	13.9999	0.0278	503.40	0.000

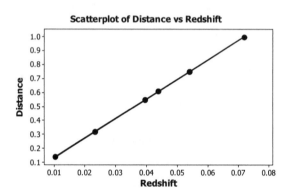

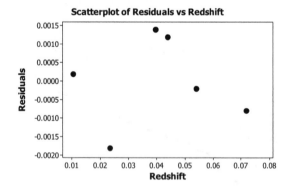

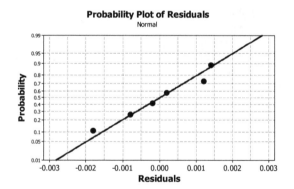

21. $\hat{y} = -157 + 40.2x$; The best predicted weight is $\hat{y} = -157 + 40.2(2) = -76.6$ kg. (Tech: -76.5 kg). That prediction is a negative weight that cannot be correct. The overhead width of 2 cm is well beyond the scope of the available sample widths, so the extrapolation might be off by a considerable amount.

MINITAB

Predictor	Coef	SE Coef	T	P
Constant	-156.88	57.41	-2.73	0.052
Width	40.182	6.712	5.99	0.004

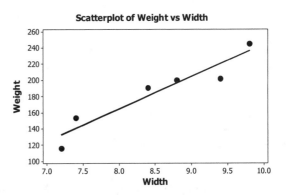

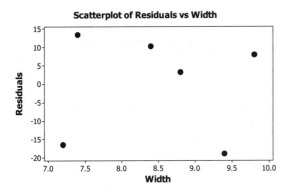

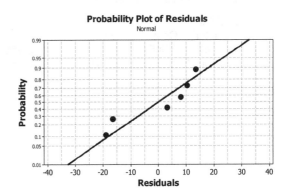

23. $\hat{y} = 43.6 + 1.31x$; The regression line does not fit the data well, so the best predicted value is $\bar{y} = 163.2$ mm Hg.

MINITAB

Predictor	Coef	SE Coef	T	P
Constant	43.56	39.93	1.09	0.355
Right Arm	1.3147	0.4361	3.01	0.057

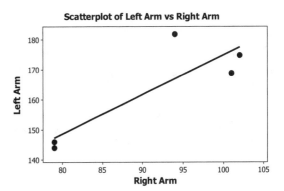

23. (continued)

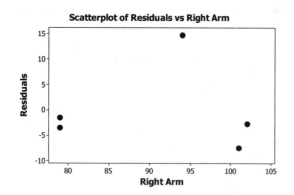

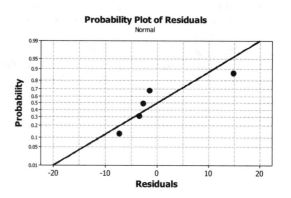

25. $\hat{y} = 2.57 + 0.172x$; The regression line does not fit the data well, so the best predicted value is $\bar{y} = \$3.05$. The predicted price is not very close to the actual price of $2.93.

MINITAB

Predictor	Coef	SE Coef	T	P
Constant	2.5662	0.9732	2.64	0.039
Regular	0.1718	0.3491	0.49	0.640

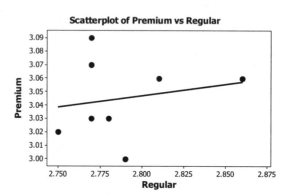

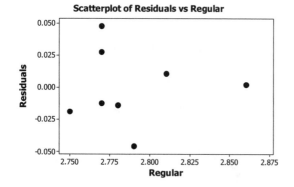

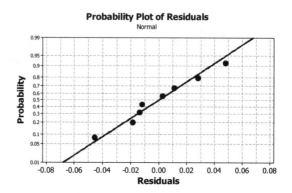

27. $\hat{y} = -0.00396 + 3.14x$; The best predicted value is $\hat{y} = -0.00396 + 3.14(1.50) = 4.7$ cm. Even though the diameter of 1.50 cm is beyond the scope of the sample diameters, the predicted value yields the actual circumference.

MINITAB

Predictor	Coef	SE Coef	T	P
Constant	-0.00396	0.01883	-0.21	0.840
Diameter	3.14274	0.00129	2443.98	0.000

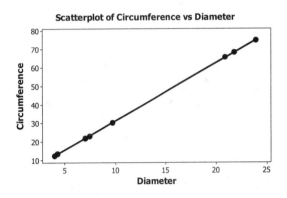

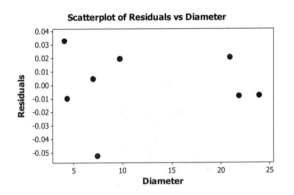

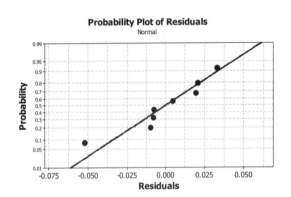

29. $\hat{y} = 109 - 0.00670x$; The regression line does not fit the data well, so the best predicted IQ score is $\overline{y} = 101$.

MINITAB

Predictor	Coef	SE Coef	T	P
Constant	108.55	28.17	3.85	0.001
VOLUME	-0.00670	0.02487	-0.27	0.791

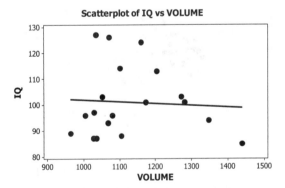

29. (continued)

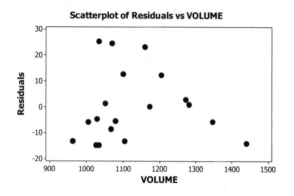

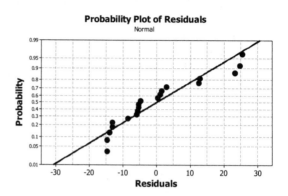

31. $\hat{y} = 13,400 + 0.302x$; The best predicted value is $\hat{y} = 13,400 + 0.302(10,000) = 16,400$ (Tech: 16,458).

MINITAB

Predictor	Coef	SE Coef	T	P
Constant	13439	2239	6.00	0.000
M1	0.3019	0.1222	2.47	0.017

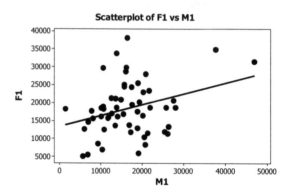

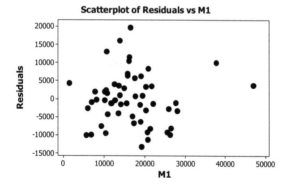

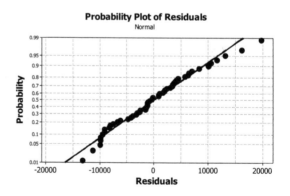

33. With $\beta_1 = 0$, the regression line is horizontal so that different values of x result in the same y value, and there is no correlation between x and y.

Section 10-4

1. The value of $s_e = 17.5436$ cm is the standard error of estimate, which is a measure of the differences between the observed weights and the weights predicted from the regression equation. It is a measure of the variation of the sample points about the regression line.

3. The coefficient of determination is $r^2 = (0.356)^2 = 0.127$. We know that 12.7% of the variation in weight is explained by the linear correlation between height and weight, and 87.3% of the variation in weight is explained by other factors and/or random variation.

5. $r^2 = (0.933)^2 = 0.870$. 87.0% of the variation in waist size is explained by the linear correlation between weight and waist size, and 13.0% of the variation in waist size is explained by other factors and/or random variation.

7. $r^2 = (-0.793)^2 = 0.629$. 62.9% of the variation in highway fuel consumption is explained by the linear correlation between weight and highway fuel consumption, and 37.1% of the variation in highway fuel consumption is explained by other factors and/or random variation.

9. $r = 0.842$. Critical values: $r = \pm 0.312$ (assuming a 0.05 significance level). P-value = 0.000. There is sufficient evidence to support a claim of a linear correlation between foot length and height.

11. $\hat{y} = 64.1 + 4.29(29.0) = 189$ cm

13. $160 \text{ cm} < y < 183 \text{ cm}$

$$\hat{y} = 64.1 + 4.29(25) = 171.35$$

$$E = t_{\alpha/2} s_e \sqrt{1 + \frac{1}{n} + \frac{n(x_0 - \bar{x})^2}{n(\Sigma x^2) - (\Sigma x)^2}} = 2.024(5.50571)\sqrt{1 + \frac{1}{40} + \frac{40(25 - 25.68)^2}{40(26530.92) - (1027.2)^2}} = 11.299$$

15. $149 \text{ cm} < y < 168 \text{ cm}$

$$\hat{y} = 64.1 + 4.29(22) = 158.48$$

$$E = t_{\alpha/2} s_e \sqrt{1 + \frac{1}{n} + \frac{n(x_0 - \bar{x})^2}{n(\Sigma x^2) - (\Sigma x)^2}} = 1.686(5.50571)\sqrt{1 + \frac{1}{40} + \frac{40(22 - 25.68)^2}{40(26530.92) - (1027.2)^2}} = 9.797$$

17. a. 10,626.59

     b. 68.83577

     c. $38.0°F < y < 60.4°F$

```
MINITAB
Analysis of Variance
Source          DF      SS      MS       F       P
Regression       1    10627   10627   771.88   0.000
Residual Error   5      69      14
Total            6    10695

Predicted Values for New Observations
Obs     Fit    SE Fit   95% CI          95% PI
1      49.19   2.31    (43.26, 55.12)   (37.96, 60.42)
```

19. a.   0.466276   b.   0.000007359976

c.   0.168 billion light-years $< y <$ 0.176 billion light-years

MINITAB
Analysis of Variance

Source	DF	SS	MS	F	P
Regression	1	0.46628	0.46628	253411.69	0.000
Residual Error	4	0.00001	0.00000		
Total	5	0.46628			

Predicted Values for New Observations

Obs	Fit	SE Fit	90% CI	90% PI
1	0.17200	0.00095	(0.16997, 0.17403)	(0.16847, 0.17554)

21. $58.9 < \beta_0 < 103$; $2.46 < \beta_1 < 3.98$

CI for β_0

$$b_0 - E < \beta_0 < b_0 + E$$

$$b_0 = 80.93; \ E = t_{\alpha/2} s_e \sqrt{\frac{1}{n} + \frac{\bar{x}^2}{\Sigma x^2 - \frac{(\Sigma x)^2}{n}}} = 2.024(5.94376)\sqrt{\frac{1}{40} + \frac{29.02^2}{33933 - \frac{1160.7^2}{40}}} = 22.06$$

CI for β_1

$$b_1 - E < \beta_1 < b_1 + E$$

$$b_1 = 3.2186; \ E = t_{\alpha/2} \cdot \frac{s_e}{\sqrt{\Sigma x^2 - \frac{(\Sigma x)^2}{n}}} = 2.024 \cdot \frac{5.94376}{\sqrt{33933 - \frac{1160.7^2}{40}}} = 0.757$$

Section 10-5

1.   The response variable is weight and the predictor variables are length and chest size.

3.   The unadjusted R^2 increases (or remains the same) as more variables are included, but the adjusted R^2 is adjusted for the number of variables and sample size. The unadjusted R^2 incorrectly suggests that the best multiple regression equation is obtained by including all of the available variables, but by taking into account the sample size and number of predictor variables, the adjusted R^2 is much more helpful in weeding out variables that should not be included.

5.   LDL = 47.4 + 0.085 WT + 0.497 SYS.

7.   No. The P-value of 0.149 is not very low, and the values of R^2 (0.098) and adjusted R^2 (0.049) are not high. Although the multiple regression equation fits the sample data best, it is not a good fit.

9.   HWY (highway fuel consumption) because it has the best combination of small P-value (0.000) and highest adjusted R^2 (0.920).

11.   CITY = –3.15 + 0.819 HWY. That equation has a low P-value of 0.000 and its adjusted R^2 value of 0.920 isn't very much less than the values of 0.928 and 0.935 that use two predictor variables, so in this case it is better to use the one predictor variable instead of two.

13. The best regression equation is $\hat{y} = 0.127 + 0.0878x_1 - 0.0250x_2$, where x_1 represents tar and x_2 represents carbon monoxide. It is best because it has the highest adjusted R^2 value of 0.927 and the lowest P-value of 0.000. It is a good regression equation for predicting nicotine content because it has a high value of adjusted R^2 and a low P-value.

```
MINITAB
Predictor         Coef    SE Coef       T      P
Constant       0.08000    0.06611    1.21   0.239
100 Tar       0.063333  0.004832   13.11   0.000
S = 0.0869783  R-Sq = 88.2%  R-Sq(adj) = 87.7%


Predictor         Coef    SE Coef       T      P
Constant        0.3281     0.1378    2.38   0.026
100 CO        0.039721   0.008967    4.43   0.000
S = 0.185937  R-Sq = 46.0%  R-Sq(adj) = 43.7%


Predictor         Coef    SE Coef       T      P
Constant       0.12714    0.05230    2.43   0.024
100 Tar       0.087797   0.007062   12.43   0.000
100 CO       -0.025004   0.006130   -4.08   0.000
S = 0.0671065  R-Sq = 93.3%  R-Sq(adj) = 92.7%
```

15. The best regression equation is $\hat{y} = 109 - 0.00670x_1$, where x_1 represents volume. It is best because it has the highest adjusted R^2 value of -0.0513 and the lowest P-value of 0.791. The three regression equations all have adjusted values of R^2 that are very close to 0, so none of them are good for predicting IQ. It does not appear that people with larger brains have higher IQ scores.

```
MINITAB
Predictor       Coef    SE Coef       T      P
Constant      108.55      28.17    3.85   0.001
VOL          -0.00670    0.02487   -0.27   0.791
S = 13.5455  R-Sq = 0.4%  R-Sq(adj) = 0.0%


Predictor       Coef    SE Coef       T      P
Constant      101.14      12.46    8.11   0.000
WT           -0.0018      0.1554   -0.01   0.991
S = 13.5728  R-Sq = 0.0%  R-Sq(adj) = 0.0%


Predictor       Coef    SE Coef       T      P
Constant      108.26      29.72    3.64   0.002
VOL          -0.00694    0.02616   -0.27   0.794
WT            0.0072      0.1631    0.04   0.965
S = 13.9375  R-Sq = 0.4%  R-Sq(adj) = 0.0%
```

17. For H_0: $\beta_1 = 0$, the test statistic is $t = \dfrac{0.7072}{0.1289} = 5.486$, the P-value is 0.000, and the critical values are $t = \pm 2.110$, so reject H_0 and conclude that the regression coefficient of $b_1 = 0.707$ should be kept. For H_0: $\beta_2 = 0$, the test statistic is $t = \dfrac{0.1636}{0.1266} = 1.292$, the P-value is 0.213, and the critical values are $t = \pm 2.110$, so fail to reject H_0 and conclude that the regression coefficient of $b_2 = 0.164$ should be omitted. It appears that the regression equation should include the height of the mother as a predictor variable, but the height of the father should be omitted.

19. $\hat{y} = 3.06 + 82.4x_1 + 2.91x_2$, where x_1 represents sex and x_2 represents age.

Female: $\hat{y} = 3.06 + 82.4(0) + 2.91(20) = 61$ lb ; male: $\hat{y} = 3.06 + 82.4(1) + 2.91(20) = 144$ lb . The sex of the bear does appear to have an effect on its weight. The regression equation indicates that the predicted weight of a male bear is about 82 lb more than the predicted weight of a female bear with other characteristics being the same.

19. (continued)

MINITAB

Predictor	Coef	SE Coef	T	P
Constant	3.06	22.46	0.14	0.892
SEX	82.38	20.80	3.96	0.000
AGE	2.9053	0.2974	9.77	0.000

Section 10-6

1. Since the area of a square is the square of its side, the best model is $y = x^2$; quadratic; $R^2 = 1$

3. 10.3% of the variation in Super Bowl points can be explained by the quadratic model that relates the variable of year and the variable of points scored. Because such a small percentage of the variation is explained by the model, the model is not very useful.

5. Quadratic: $d = -4.88t^2 + 0.0214t + 300$

Model	R^2
Linear	0.962
Quadratic	1.000
Logarithmic	0.831
Exponential	0.933
Power	0.783

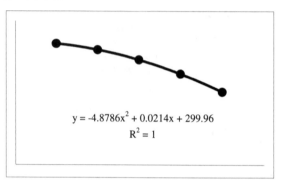

7. Exponential: $y = 100(1.03^x)$ The value of R^2 is slightly higher for the exponential model.

Model	R^2
Linear	0.999
Quadratic	1.000
Logarithmic	0.900
Exponential	0.999
Power	0.918

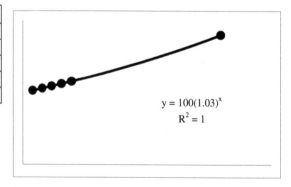

9. Power: $y = 65.7x^{-0.945}$. Prediction for the 22nd day: $y = 65.7(22)^{-0.945} = \$3.5$ million, which isn't very close to the actual amount of $2.2 million. The model does not take into account the fact that movies do better on weekend days.

Model	R^2
Linear	0.562
Quadratic	0.774
Logarithmic	0.820
Exponential	0.792
Power	0.842

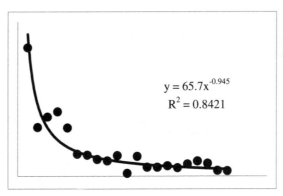

11. Logarithmic: $y = 3.22 + 0.293 \ln x$

Model	R^2
Linear	0.620
Quadratic	0.901
Logarithmic	0.997
Exponential	0.566
Power	0.989

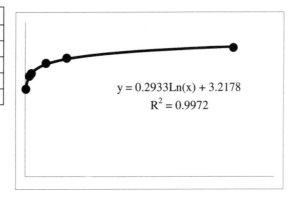

13. Exponential: $y = 10(2^x)$

Model	R^2
Linear	0.771
Quadratic	0.975
Logarithmic	0.549
Exponential	1.000
Power	0.927

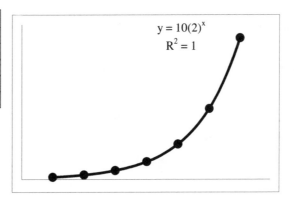

15. Quadratic: $y = 125x^2 - 439x + 3438$. The projected value for 2010 is $y = 125(21)^2 - 439(21) + 3438$
$= 49,344$ (Tech: 49,312), which is dramatically greater than the actual value of 11,655.

Model	R^2
Linear	0.893
Quadratic	0.995
Logarithmic	0.657
Exponential	0.958
Power	0.767

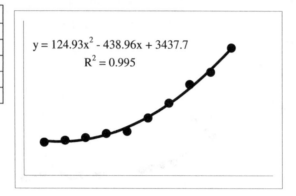

17. a.   Exponential: $y = 2^{\frac{2}{3}(x-1)}$ [or $y = 0.629961(1.587401)^x$ for an initial value of 1 that doubles every 1.5
years].

b.   Exponential: $y = 1.36558(1.42774)^x$, where 1971 is coded as 1.

Model	R^2
Linear	0.380
Quadratic	0.55
Logarithmic	0.158
Exponential	0.990
Power	0.790

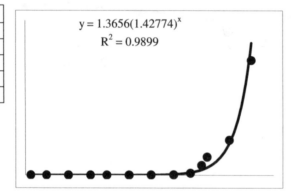

c.   Moore's law does appear to be working reasonably well. With $R^2 = 0.990$, the model appears to be
very good.

Chapter Quick Quiz

1. $r = \pm 0.878$

2.   Based on the critical values of ± 0.878 (assuming a 0.05 significance level), conclude that there is not
sufficient evidence to support the claim of a linear correlation between systolic and diastolic readings.

3.   The best predicted diastolic reading is 90.6, which is the mean of the five sample diastolic readings.

4.   The best predicted diastolic reading is $\hat{y} = -1.99 + 0.698(125) = 85.3$, which is found by substituting 125
for x in the regression equation.

5. $r^2 = 0.342$

6.   False; there could be another relationship.   7.   False, correlation does not imply causation.

8. $r = 1$

9.   Because r must be between −1 and 1 inclusive, the value of 3.335 is the result of an error in the
calculations.

10. $r = -1$

Review Exercises

1. a. $r = 0.926$. Critical values: $r = \pm 0.707$ (assuming a 0.05 significance level). P-value $= 0.001$. There is sufficient evidence to support the claim that there is a linear correlation between duration and interval-after time.

      MINITAB
      Pearson correlation of After and Duration = 0.926
      P-Value = 0.001

  b. $r^2 = (0.926)^2 = 0.857 = 85.7\%$

  c. $\hat{y} = 34.8 + 0.234x$

MINITAB

Predictor	Coef	SE Coef	T	P
Constant	34.770	8.732	3.98	0.007
Duration	0.23406	0.03908	5.99	0.001

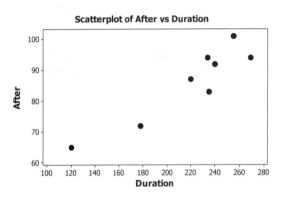

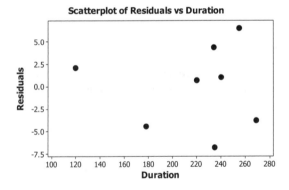

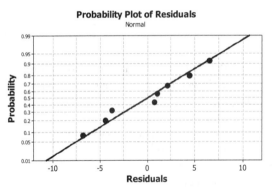

  d. $\hat{y} = 34.8 + 0.234(200) = 81.6$ min

2. a.   The scatterplot suggests that there is not sufficient sample evidence to support the claim of a linear correlation between heights of eruptions and interval-after times.

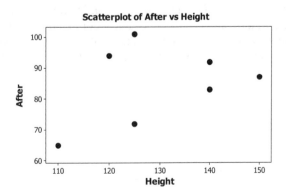

2.   (continued)

b. $r = 0.269$. Critical values: $r = \pm 0.707$ (assuming a 0.05 significance level). *P*-value = 0.519. There is not sufficient evidence to support the claim that there is a linear correlation between height and interval-after time.

> MINITAB
> Pearson correlation of Height and After = 0.269
> P-Value = 0.519

c. $\hat{y} = 54.3 + 0.246x$

> MINITAB
>
> | Predictor | Coef | SE Coef | T | P |
> |-----------|------|---------|------|-------|
> | Constant | 54.27 | 46.53 | 1.17 | 0.288 |
> | Height | 0.2465 | 0.3597 | 0.69 | 0.519 |

d. $\hat{y} = 54.3 + 0.246(100) = 78.9$ min

3.   a.   The scatterplot suggests that there is not sufficient sample evidence to support the claim of a linear correlation between duration and height.

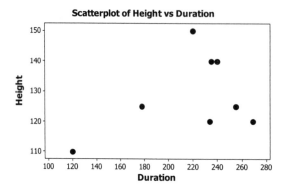

b. $r = 0.389$. Critical values: $r = \pm 0.707$ (assuming a 0.05 significance level). *P*-value = 0.340. There is not sufficient evidence to support the claim that there is a linear correlation between duration and height.

> MINITAB
> Pearson correlation of Height and Duration = 0.389
> P-Value = 0.340

c. $\hat{y} = 105 + 0.108x$

> MINITAB
>
> | Predictor | Coef | SE Coef | T | P |
> |-----------|--------|---------|------|-------|
> | Constant | 105.19 | 23.22 | 4.53 | 0.004 |
> | Duration | 0.1076 | 0.1039 | 1.04 | 0.340 |

d.   The regression line does not fit the points well, so the best predicted height $\bar{y} = 128.8$ ft.

4. $r = 0.450$. Critical values: $r = \pm 0.632$ (assuming a 0.05 significance level). *P*-value = 0.192. There is not sufficient evidence to support the claim that there is a linear correlation between time and height. Although there is no linear correlation between time and height, the scatterplot shows a very distinct pattern revealing that time and height are associated by some function that is not linear.

4.   (continued)

MINITAB
Pearson correlation of Height(m) and Time(sec)
   = 0.450
P-Value = 0.192

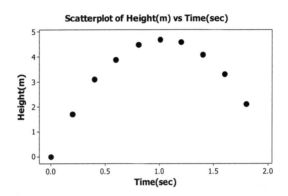

5.   AFTER = 50.1 + 0.242 Duration − 0.178 BEFORE, or $\hat{y} = 50.1 + 0.242x_1 - 0.178x_2$. $R^2 = 0.872$; adjusted $R^2 = 0.820$; P-value = 0.006. With high values of R^2 and adjusted R^2 and a small P-value of 0.006, it appears that the regression equation can be used to predict the time interval after an eruption given the duration of the eruption and the time interval before that eruption.

```
MINITAB
Predictor        Coef   SE Coef       T       P
Constant        50.09     22.07    2.27   0.072
Duration      0.24179   0.04177    5.79   0.002
Before     -   0.1779    0.2336   -0.76   0.481
S = 5.15785   R-Sq = 87.2%   R-Sq(adj) = 82.0%
```

Cumulative Review Exercises

1. $\bar{x} = 3.3$ lb , $s = 5.7$ lb

2.   The highest weight before the diet is 212 lb, which converts to $z = \dfrac{212 - 179.4}{21.0} = 1.55$. The highest weight is not unusual because its z score of 1.55 shows that it is within 2 standard deviations of the mean.

3. $H_0 : \mu_d = 0$. $H_0 : \mu_d > 0$. Test statistic: $t = 1.613$. Critical value: $t = 1.895$. P-value > 0.05 (Tech: 0.075). Fail to reject H_0. There is not sufficient evidence to support the claim that the diet is effective.

```
MINITAB
Paired T for Before - After
95% lower bound for mean difference: -0.57
T-Test of mean difference = 0 (vs > 0): T-Value = 1.61  P-Value = 0.075
```

4.   161.8 lb $< \mu <$ 197.0 lb. We have 95% confidence that the interval limits of 161.8 lb and 197.0 lb contain the true value of the mean of the population of all subjects before the diet.

```
MINITAB
Variable  N    Mean StDev SE Mean      95% CI
Before    8  179.38 21.04     7.44  (161.79, 196.96)
```

5.   a. $r = 0.965$. Critical values: $r = \pm 0.707$ (assuming a 0.05 significance level). P-value = 0.000. There is sufficient evidence to support the claim that there is a linear correlation between before and after weights.

```
MINITAB
Pearson correlation of Before and After = 0.965
P-Value = 0.000
```

b. $r = 1$ c. $r = 1$

d.   The effectiveness of the diet is determined by the amounts of weight lost, but the linear correlation coefficient is not sensitive to different amounts of weight loss. Correlation is not a suitable tool for testing the effectiveness of the diet.

6. a. $z = \dfrac{3500 - 3420}{495} = 0.162;\ \ P(z > 0.162) = 43.64\%.$ (Tech: 43.58%)

    b.   10th percentile: $x = \mu + z \cdot \sigma = 3420 - 1.28 \cdot 495 = 2786.4$ g (Tech: 2785.6 g)

    c. $z = \dfrac{2450 - 3420}{495} = -1.96;\ P(z < -1.96) = 0.0250.$

         $z = \dfrac{4390 - 3420}{495} = 1.96;\ P(z > 1.96) = 0.0250.$

         $0.0250 + 0.0250 = 0.0500 = 5.00\%.$ Yes, many of the babies do require special treatment.

7. a. H_0: $p = 0.5.$ H_1: $p > 0.5.$ Test statistic: $z = 3.84$. Critical value: $z = 1.645$. P-value: 0.0001. Reject H_0.
         There is sufficient evidence to support the claim that the majority of us say that honesty is always the
         best policy.

         MINITAB
         Test of p = 0.5 vs p > 0.5

Sample	X	N	Sample p	95% Lower Bound	Z-Value	P-Value
1	269	456	0.589912	0.552026	3.84	0.000

    b.   The sample is a voluntary response (or self-selected) sample. This type of sample suggests that the
         results given in part (a) are not necessarily valid.

8. a.   Nominal

    b.   Ratio

    c.   Discrete

    d. $\dfrac{304}{529} = 0.575$

    e.   Parameter

9. a. $\left(\dfrac{304}{529}\right)^2 = 0.330$

    b. $\dfrac{304 + 156}{529} = \dfrac{460}{529} = 0.870$

    c. $\dfrac{514}{529} = 0.972$

    d. $\dfrac{39}{529} = 0.0737 = 7.37\%$

10.

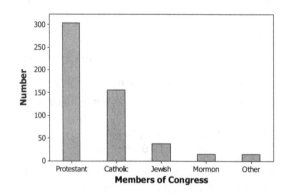

Chapter 11: Goodness-of-Fit and Contingency Tables

Section 11-2

1. The test is to determine whether the observed frequency counts agree with the claimed uniform distribution so that frequencies for the different days are equally likely.

3. Because the given frequencies differ substantially from frequencies that are all about the same, the χ^2 test statistic should be large and the P-value should be small.

5. Test statistic: $\chi^2 = 1934.979$. Critical value: $\chi^2 = 12.592$. P-value = 0.000. There is sufficient evidence to warrant rejection of the claim that the days of the week are selected with a uniform distribution with all days having the same chance of being selected.

7. Critical value: $\chi^2 = 16.919$. P-value > 0.10 (Tech: 0.516). There is not sufficient evidence to warrant rejection of the claim that the observed outcomes agree with the expected frequencies. The slot machine appears to be functioning as expected.

9. Test statistic: $\chi^2 = 10.375$. Critical value: $\chi^2 = 19.675$. P-value > 0.10 (Tech: 0.497). There is not sufficient evidence to warrant rejection of the claim that homicides in New York City are equally likely for each of the 12 months. There is not sufficient evidence to support the police commissioner's claim that homicides occur more often in the summer when the weather is better.

    MINITAB
N	DF	Chi-Sq	P-Value
512	11	10.375	0.497

11. Test statistic: $\chi^2 = 5.860$. Critical value: $\chi^2 = 11.071$. P-value > 0.10 (Tech: P-value = 0.320). There is not sufficient evidence to support the claim that the outcomes are not equally likely. The outcomes appear to be equally likely, so the loaded die does not appear to behave differently from a fair die.

    MINITAB
N	DF	Chi-Sq	P-Value
200	5	5.86	0.320

13. Test statistic: $\chi^2 = 13.483$. Critical value: $\chi^2 = 16.919$. P-value > 0.10 (Tech: 0.142). There is not sufficient evidence to warrant rejection of the claim that the likelihood of winning is the same for the different post positions. Based on these results, post position should not be considered when betting on the Kentucky Derby race.

    MINITAB
N	DF	Chi-Sq	P-Value
116	9	13.4828	0.142

15. Test statistic: $\chi^2 = 29.814$. Critical value: $\chi^2 = 16.812$. P-value < 0.005 (Tech: 0.000). There is sufficient evidence to warrant rejection of the claim that the different days of the week have the same frequencies of police calls. The highest numbers of calls appear to fall on Friday and Saturday, and these are weekend days with disproportionately more partying and drinking.

    MINITAB
N	DF	Chi-Sq	P-Value
1095	6	29.8137	0.000

17. Test statistic: $\chi^2 = 7.579$. Critical value: $\chi^2 = 7.815$. P-value > 0.05 (Tech: 0.056). There is not sufficient evidence to warrant rejection of the claim that the actual numbers of games fit the distribution indicated by the proportions listed in the given table.

17. (continued)

Games Played	O	E	$O-E$	$(O-E)^2$	$\dfrac{(O-E)^2}{E}$
4	20	$103 \cdot 0.125 = 12.875$	7.125	50.76563	3.942961
5	23	$103 \cdot 0.2500 = 25.75$	−2.75	7.5625	0.293689
6	23	$103 \cdot 0.3125 = 32.1875$	−9.1875	84.41016	2.622451
7	37	$103 \cdot 0.3125 = 32.1875$	4.8125	23.16016	0.719539
				Sum	7.578641

19. Test statistic: $\chi^2 = 6.682$. Critical value: $\chi^2 = 11.071$ (assuming a 0.05 significance level). *P*-value > 0.10 (Tech: 0.245). There is not sufficient evidence to warrant rejection of the claim that the color distribution is as claimed.

Color	O	E	$O-E$	$(O-E)^2$	$\dfrac{(O-E)^2}{E}$
Red	13	$100 \cdot 0.13 = 13$	0	0	0
Orange	25	$100 \cdot 0.20 = 20$	5	25	1.25
Yellow	8	$100 \cdot 0.14 = 14$	−6	36	2.571429
Brown	8	$100 \cdot 0.13 = 13$	−5	25	1.923077
Blue	27	$100 \cdot 0.24 = 24$	3	9	0.375
Green	19	$100 \cdot 0.16 = 16$	3	9	0.5625
				Sum	6.682005

21. Test statistic: $\chi^2 = 3650.251$. Critical value: $\chi^2 = 20.090$. P-value < 0.005 (Tech: 0.000). There is sufficient evidence to warrant rejection of the claim that the leading digits are from a population with a distribution that conforms to Benford's law. It does appear that the checks are the result of fraud (although the results cannot confirm that fraud is the cause of the discrepancy between the observed results and the expected results).

```
MINITAB
N    DF  Chi-Sq    P-Value
784  8   3650.25    0.000
```

23. Test statistic: $\chi^2 = 1.762$. Critical value: $\chi^2 = 15.507$. P-value > 0.10 (Tech: 0.988). There is not sufficient evidence to warrant rejection of the claim that the leading digits are from a population with a distribution that conforms to Benford's law. The tax entries do appear to be legitimate.

```
MINITAB
N    DF  Chi-Sq    P-Value
511  8   1.76216   0.987
```

25. a. 6, 13, 15, 6

b. $z = \dfrac{155.41 - 162}{6.595} = -1;\ P(z < -1) = 0.1587\,;$

$z = \dfrac{162.005 - 162}{6.595} = 0;\ P(-1 < z < 0) = 0.5000 - 0.1587 = 0.3413\,;$

$z = \dfrac{168.601 - 162}{6.595} = 1;\ P(0 < z < 1) = 0.8413 - 0.5000 = 0.3413\,:$

$z = \dfrac{215 - 280}{65} = -1;\ P(z > 1) = 0.1587$

(Tech: 0.1587, 0.3413, 0.3414, 0.1586)

25. (continued)

c. $40 \cdot 0.1587 = 6.348$, $40 \cdot 0.3413 = 13.652$, , $40 \cdot 0.3413 = 13.652$, $40 \cdot 0.1587 = 6.348$

(Tech: 6.348, 13.652, 13.656, 6.344)

d.   Test statistic: $\chi^2 = 0.202$ (Tech: 0.201). Critical value: $\chi^2 = 11.345$. P-value > 0.10 (Tech: 0.977). There is not sufficient evidence to warrant rejection of the claim that heights were randomly selected from a normally distributed population. The test suggests that the data are from a normally distributed population.

Height	O	E	$O-E$	$(O-E)^2$	$\dfrac{(O-E)^2}{E}$
Less than 155.410	6	6.348	−0.348	0.121104	0.019078
155.410–162.005	13	13.652	−0.652	0.425104	0.031139
162.005–168.601	15	13.652	1.348	1.817104	0.133102
Greater than 168.601	6	6.348	−0.348	0.121104	0.019078

Sum   0.202395

```
MINITAB
N    DF   Chi-Sq      P-Value
40   3    0.202395    0.977
```

Section 11-3

1.   Because the P-value of 0.216 is not small (such as 0.05 or lower), fail to reject the null hypothesis of independence between the treatment and whether the subject stops smoking. This suggests that the choice of treatment doesn't appear to make much of a difference.

3. $df = (3-1)(2-1) = 2$ and the critical value is $\chi^2 = 5.991$.

5.   Test statistic: $\chi^2 = 3.409$. Critical value: $\chi^2 = 3.841$. P-value > 0.05 (Tech: 0.0648). There is not sufficient evidence to warrant rejection of the claim that the form of the 100-Yuan gift is independent of whether the money was spent. There is not sufficient evidence to support the claim of a denomination effect.

7.   Test statistic: $\chi^2 = 25.571$. Critical value: $\chi^2 = 3.841$. P-value < 0.005 (Tech: 0.000). There is sufficient evidence to warrant rejection of the claim that whether a subject lies is independent of the polygraph test indication. The results suggest that polygraphs are effective in distinguishing between truths and lies, but there are many false positives and false negatives, so they are not highly reliable.

```
MINITAB
Expected counts are printed below observed counts

            No (Did      Yes
            Not Lie)    (Lied)     Total
    1          15          42        57
            27.34       29.66
    2          32           9        41
            19.66       21.34
Total         47          51        98

Chi-Sq = 25.571, DF = 1, P-Value = 0.000
```

9. Test statistic: $\chi^2 = 42.557$. Critical value: $\chi^2 = 3.841$. P-value < 0.005 (Tech: 0.000). There is sufficient evidence to warrant rejection of the claim that the sentence is independent of the plea. The results encourage pleas for guilty defendants.

MINITAB

Expected counts are printed below observed counts

	Guilty Plea	Not Guilty Plea	Total
1	392	58	450
	418.48	31.52	
2	564	14	578
	537.52	40.48	
Total	956	72	1028

Chi-Sq = 42.557, DF = 1, P-Value = 0.000

11. Test statistic: $\chi^2 = 0.164$. Critical value: $\chi^2 = 3.841$. P-value > 0.10 (Tech: 0.686). There is not sufficient evidence to warrant rejection of the claim that the gender of the tennis player is independent of whether the call is overturned.

MINITAB

Expected counts are printed below observed counts

	Yes	No	Total
1	421	991	1412
	416.90	995.10	
2	220	539	759
	224.10	534.90	
Total	641	1530	2171

Chi-Sq = 0.164, DF = 1, P-Value = 0.686

13. Test statistic: $\chi^2 = 14.589$. Critical value: $\chi^2 = 9.488$. P-value < 0.01 (Tech: 0.0056). There is sufficient evidence to warrant rejection of the claim that the direction of the kick is independent of the direction of the goalkeeper jump. The results do not support the theory that because the kicks are so fast, goalkeepers have no time to react, so the directions of their jumps are independent of the directions of the kicks.

MINITAB
Chi-Sq = 14.589, DF = 4, P-Value = 0.006

15. Test statistic: $\chi^2 = 2.925$. Critical value: $\chi^2 = 5.991$. P-value > 0.10 (Tech: 0.232). There is not sufficient evidence to warrant rejection of the claim that getting a cold is independent of the treatment group. The results suggest that echinacea is not effective for preventing colds.

MINITAB
Chi-Sq = 2.925, DF = 2, P-Value = 0.232

17. Test statistic: $\chi^2 = 20.271$. Critical value: $\chi^2 = 15.086$. P-value < 0.005 (Tech: 0.0011). There is sufficient evidence to warrant rejection of the claim that cooperation of the subject is independent of the age category. The age group of 60 and over appears to be particularly uncooperative.

MINITAB
Chi-Sq = 20.271, DF = 5, P-Value = 0.001

19. Test statistic: $\chi^2 = 0.773$. Critical value: $\chi^2 = 11.345$. P-value > 0.10 (Tech: 0.856). There is not sufficient evidence to warrant rejection of the claim that getting an infection is independent of the treatment. The atorvastatin treatment does not appear to have an effect on infections.

MINITAB
Chi-Sq = 0.773, DF = 3, P-Value = 0.856

21. Test statistics: $\chi^2 = 12.1619258$ and $z = 3.487395274$, so that $z^2 = \chi^2$. Critical values: $\chi^2 = 3.841$ and $z^2 = \pm 1.96$, so $z^2 = \chi^2$ (approximately).

MINITAB
Expected counts are printed below observed counts

	Purchased Gum	Kept the Money	Total
1	27	16	43
	18.84	24.16	
2	12	34	46
	20.16	25.84	
Total	39	50	89

Chi-Sq = 12.162, DF = 1, P-Value = 0.000

MINITAB
Difference = p (1) - p (2)
Estimate for difference: 0.372308
95% CI for difference: (0.178143, 0.566473)
Test for difference = 0 (vs not = 0): Z = 3.49 P-Value = 0.000

Chapter Quick Quiz

1. H_0: $p_1 = p_2 = p_3 = p_4 = p_5$. H_1: At least one of the probabilities is different from the others.

2. $O = 23$ and $E = \dfrac{107}{5} = 21.4$.

3. Right-tailed.

4. df = 4 and the critical value is $\chi^2 = 9.488$.

5. There is not sufficient evidence to warrant rejection of the claim that occupation injuries occur with equal frequency on the different days of the week.

6. H_0: Response to the question is independent of gender. H_1: Response to the question and gender are dependent.

7. Chi-square distribution.

8. Right-tailed.

9. $df = (2-1)(3-1) = 2$ and the critical value is $\chi^2 = 5.991$.

10. There is not sufficient evidence to warrant rejection of the claim that response is independent of gender.

Review Exercises

1. Test statistic: $\chi^2 = 931.347$. Critical value: $\chi^2 = 16.812$. P-value: 0.000. There is sufficient evidence to warrant rejection of the claim that auto fatalities occur on the different days of the week with the same frequency. Because people generally have more free time on weekends and more drinking occurs on weekends, the days of Friday, Saturday, and Sunday appear to have disproportionately more fatalities.

2. Test statistic: $\chi^2 = 6.500$. Critical value: $\chi^2 = 16.919$. P-value > 0.10 (Tech: 0.689). There is not sufficient evidence to warrant rejection of the claim that the last digits of 0, 1, 2, . . . , 9 occur with the same frequency. It does appear that the weights were obtained through measurements.

MINITAB
N	DF	Chi-Sq	P-Value
80	9	6.5	0.689

3. Test statistic: $\chi^2 = 288.448$. Critical value: $\chi^2 = 24.725$. P-value < 0.005 (Tech: 0.000). There is sufficient evidence to warrant rejection of the claim that weather-related deaths occur in the different months with the same frequency. The summer months appear to have disproportionately more weather-related deaths, and that is probably due to the fact that vacations and outdoor activities are much greater during those months.

    MINITAB
    ```
    N    DF  Chi-Sq   P-Value
    489  11  288.448  0.000
    ```

4. Test statistic: $\chi^2 = 10.708$. Critical value: $\chi^2 = 3.841$. P-value: 0.00107. There is sufficient evidence to warrant rejection of the claim that wearing a helmet has no effect on whether facial injuries are received. It does appear that a helmet is helpful in preventing facial injuries in a crash.

5. Test statistic: $\chi^2 = 4.955$. Critical value: $\chi^2 = 3.841$. P-value < 0.05 (Tech: 0.0260). There is sufficient evidence to warrant rejection of the claim that when flipping or spinning a penny, the outcome is independent of whether the penny was flipped or spun. It appears that the outcome is affected by whether the penny is flipped or spun. If the significance level is changed to 0.01, the critical value changes to 6.635, and we fail to reject the given claim, so the conclusion does change. All expected counts are greater than 5.

    Expected counts are printed below observed counts
    Chi-Square contributions are printed below expected counts

    ```
                Heads     Tails   Total
       1        2048      1992    4040
              2007.29   2032.71
       2         953      1047    2000
               993.71   1006.29
    Total       3001      3039    6040
    ```

    Chi-Sq = 4.955, DF = 1, P-Value = 0.026

6. Test statistic: $\chi^2 = 4.737$. Critical value: $\chi^2 = 7.815$. P-value > 0.10 (Tech: 0.192). There is not sufficient evidence to warrant rejection of the claim that home/visitor wins are independent of the sport.

    MINITAB
    Expected counts are printed below observed counts
    Chi-Square contributions are printed below expected counts

    ```
            Basketball Baseball Hockey Football Total
       1        127        53     50       57    287
              115.97      58.57  54.47    57.99
       2         71        47     43       42    203
               82.03      41.43  38.53    41.01
    Total       198       100     93       99    490
    ```

    Chi-Sq = 4.737, DF = 3, P-Value = 0.192

Cumulative Review Exercises

1. H_0: $p = 0.5$. H_1: $p \neq 0.5$. Test statistic: $z = 7.28$. Critical values: $z = \pm 1.96$. P-value: 0.0002 (Tech: 0.0000). Reject H_0. There is sufficient evidence to warrant rejection of the claim that among those who die in weather-related deaths, the percentage of males is equal to 50%.

    MINITAB
    Test of p = 0.5 vs p not = 0.5

    ```
                                                    Exact
    Sample  X    N    Sample p   95% CI              P-Value
    1       325  489  0.664622   (0.620854, 0.706384)  0.000
    ```

2. $59.0\% < p < 65.0\%$. Because the confidence interval does not include 50% (or "half"), we should reject the stated claim.

```
MINITAB
Sample  X    N     Sample p    95% CI
1       620  1000  0.620000    (0.589098, 0.650193)
```

3. $\bar{x} = 53.7$ years, median = 60.0 years, $s = 16.1$ years. Because an age of 16 differs from the mean by more than 2 standard deviations, it is an unusual age.

4. 42.2 years $< \mu < 65.2$ years. Yes, the confidence interval limits do contain the value of 65.0 years that was found from a sample of 9269 ICU patients.

```
MINITAB
Variable  N   Mean   StDev  SE Mean   95% CI
AGES      10  53.70  16.09  5.09      (42.19, 65.21)
```

5.   a. $r = -0.0458$. Critical values: $r = \pm 0.632$. P-value = 0.900. There is not sufficient evidence to support the claim that there is a linear correlation between the numbers of boats and the numbers of manatee deaths.

```
MINITAB
Pearson correlation of Boats and Manatee Deaths = -0.046
P-Value = 0.900
```

   b. $\hat{y} = 96.1 - 0.137x$

```
MINITAB
The regression equation is
Manatee Deaths = 96.1 - 0.14 Boats

Predictor  Coef    SE Coef  T      P
Constant   96.14   99.89    0.96   0.364
Boats      -0.137  1.053    -0.13  0.900
```

   c. $\hat{y} = 96.1 - 0.137(84) = 84.6$ manatee deaths (the value of y). The predicted value is not very accurate because it is not very close to the actual value of 78 manatee deaths.

6.   a.   5th percentile: $x = \mu + z \cdot \sigma = 686 - 1.645 \cdot 34 = 630$ mm

   b. $z = \dfrac{650 - 686}{34} = -1.06$ and $P(z < -1.06) = 14.46\%$ (Tech: 14.48%). That percentage is too high, because too many women would not be accommodated.

   c. $z = \dfrac{680 - 686}{34/\sqrt{16}} = -0.706$ and $P(z > -0.706) = 76.11\%$ (Tech: 0.7599). Groups of 16 women do not occupy a cockpit; because individual women occupy the cockpit, this result has no effect on the design.

7.   a.   Statistic.

   b.   Quantitative.

   c.   Discrete.

   d.   The sampling is conducted so that all samples of the same size have the same chance of being selected.

   e.   The sample is a voluntary response sample (or self-selected sample), and those with strong feelings about the topic are more likely to respond, so it is not a valid sampling plan.

8.   a. $(0.6)^4 = 0.1296$

   b. $1 - 0.6 = 0.4$

Chapter 12: Analysis of Variance

Section 12-2

1. a. The chest deceleration measurements are categorized according to the one characteristic of size.

    b. The terminology of analysis of variance refers to the method used to test for equality of the three population means. That method is based on two different estimates of a common population variance.

3. The test statistic is $F = 3.288$, and the F distribution applies.

5. Test statistic: $F = 0.39$. P-value: 0.677. Fail to reject H_0: $\mu_1 = \mu_2 = \mu_3$. There is not sufficient evidence to warrant rejection of the claim that the three categories of blood lead level have the same mean verbal IQ score. Exposure to lead does not appear to have an effect on verbal IQ scores.

7. Test statistic: $F = 11.6102$. P-value: 0.000577. Reject H_0: $\mu_1 = \mu_2 = \mu_3$. There is sufficient evidence to warrant rejection of the claim that the three size categories have the same mean highway fuel consumption. The size of a car does appear to affect highway fuel consumption.

9. Test statistic: $F = 0.161$. P-value: 0.852. Fail to reject H_0: $\mu_1 = \mu_2 = \mu_3$. There is not sufficient evidence to warrant rejection of the claim that the three size categories have the same mean head injury measurement. The size of a car does not appear to affect head injuries.

11. Test statistic: $F = 27.2488$. P-value: 0.000. Reject H_0: $\mu_1 = \mu_2 = \mu_3$. There is sufficient evidence to warrant rejection of the claim that the three different miles have the same mean time. These data suggest that the third mile appears to take longer, and a reasonable explanation is that the third lap has a hill.

    EXCEL
    ANOVA

Source of Variation	SS	df	MS	F	P-value	F crit
Between Groups	0.103444	2	0.051722	27.24878	3.45E-05	3.885294
Within Groups	0.022778	12	0.001898			
Total	0.126222	14				

13. Test statistic: $F = 6.1413$. P-value: 0.0056. Reject H_0: $\mu_1 = \mu_2 = \mu_3 = \mu_4$. There is sufficient evidence to warrant rejection of the claim that the four treatment categories yield poplar trees with the same mean weight. Although not justified by the results from analysis of variance, the treatment of fertilizer and irrigation appears to be most effective.

    EXCEL
    ANOVA

Source of Variation	SS	df	MS	F	P-value	F crit
Between Groups	3.346455	3	1.115485	6.14127	0.005566	3.238872
Within Groups	2.9062	16	0.181638			
Total	6.252655	19				

15. Test statistic: $F = 18.9931$. *P*-value: 0.000. Reject H_0: $\mu_1 = \mu_2 = \mu_3$. There is sufficient evidence to warrant rejection of the claim that the three different types of cigarettes have the same mean amount of nicotine. Given that the king-size cigarettes have the largest mean of 1.26 mg per cigarette, compared to the other means of 0.87 mg per cigarette and 0.92 mg per cigarette, it appears that the filters do make a difference (although this conclusion is not justified by the results from analysis of variance).

EXCEL
ANOVA

Source of Variation	SS	df	MS	F	P-value	F crit
Between Groups	2.208267	2	1.104133	18.99312	2.38E-07	3.123907
Within Groups	4.1856	72	0.058133			
Total	6.393867	74				

17. The Tukey test results show different *P*-values, but they are not dramatically different. The Tukey results suggest the same conclusions as the Bonferroni test.

Section 12-3

1. The load values are categorized using two different factors of (1) femur (left or right) and (2) size of car (small, midsize, large).

3. An interaction between two factors or variables occurs if the effect of one of the factors changes for different categories of the other factor. If there is an interaction effect, we should not proceed with individual tests for effects from the row factor and column factor. If there is an interaction, we should not consider the effects of one factor without considering the effects of the other factor.

5. For interaction, the test statistic is $F = 1.72$ and the *P*-value is 0.194, so there is not sufficient evidence to conclude that there is an interaction effect. For the row variable of femur (right, left), the test statistic is $F = 1.39$ and the *P*-value is 0.246, so there is not sufficient evidence to conclude that whether the femur is right or left has an effect on measured load. For the column variable of size of the car, the test statistic is $F = 2.23$ and the *P*-value is 0.122, so there is not sufficient evidence to conclude that the car size category has an effect on the measured load.

7. For interaction, the test statistic is $F = 1.05$ and the *P*-value is 0.365, so there is not sufficient evidence to conclude that there is an interaction effect. For the row variable of sex, the test statistic is $F = 4.58$ and the *P*-value is 0.043, so there is sufficient evidence to conclude that the sex of the subject has an effect on verbal IQ score. For the column variable of blood lead level (LEAD), the test statistic is $F = 0.14$ and the *P*-value is 0.871, so there is not sufficient evidence to conclude that blood lead level has an effect on verbal IQ score. It appears that only the sex of the subject has an effect on verbal IQ score.

9. For interaction, the test statistic is $F = 3.7332$ and the *P*-value is 0.0291, so there is sufficient evidence to conclude that there is an interaction effect. The measures of self-esteem appear to be affected by an interaction between the self-esteem of the subject and the self-esteem of the target. Because there appears to be an interaction effect, we should not proceed with individual tests of the row factor (target's self-esteem) and the column factor (subject's self-esteem).

EXCEL
ANOVA

Source of Variation	SS	df	MS	F	P-value	F crit
Sample	4.5	1	4.5	4.977654	0.029079736	3.986269
Columns	2.861111	2	1.430556	1.582402	0.213176735	3.135918
Interaction	6.75	2	3.375	3.73324	0.029107515	3.135918
Within	59.66667	66	0.90404			
Total	73.77778	71				

11. a.   Test statistics and *P*-values do not change.

    b.   Test statistics and *P*-values do not change.

    c.   Test statistics and *P*-values do not change.

    d.   An outlier can dramatically affect and change test statistics and *P*-values.

Chapter Quick Quiz

1. H_0: $\mu_1 = \mu_2 = \mu_3$. Because the displayed *P*-value of 0.000 is small, reject H_0.

2.   No. Because we reject the null hypothesis of equal means, it appears that the three different power sources do not produce the same mean voltage level, so we cannot expect electrical appliances to behave the same way when run from the three different power sources.

3.   Right-tailed.

4.   Test statistic: $F = 183.01$. In general, larger test statistics result in smaller *P*-values.

5.   The sample voltage measurements are categorized using only one factor: the source of the voltage.

6.   Test a null hypothesis that three or more samples are from populations with equal means.

7.   With one-way analysis of variance, the different samples are categorized using only one factor, but with two-way analysis of variance, the sample data are categorized into different cells determined by two different factors.

8.   For interaction, the test statistic is $F = 0.19$ and the *P*-value is 0.832. Fail to reject the null hypothesis of no interaction. There does not appear to be an effect due to an interaction between sex and major.

9.   The test statistic is $F = 0.78$ and the *P*-value is 0.395. There is not sufficient evidence to support a claim that the length estimates are affected by the sex of the subject.

10. The test statistic is $F = 0.13$ and the *P*-value is 0.876. There is not sufficient evidence to support a claim that the length estimates are affected by the subject's major.

Review Exercises

1. H_0: $\mu_1 = \mu_2 = \mu_3$. Test statistic: $F = 10.10$. *P*-value: 0.001. Reject the null hypothesis. There is sufficient evidence to warrant rejection of the claim that 4-cylinder cars, 6-cylinder cars, and 8-cylinder cars have the same mean highway fuel consumption amount.

2.   For interaction, the test statistic is $F = 0.17$ and the *P*-value is 0.915, so there is not sufficient evidence to conclude that there is an interaction effect. For the row variable of site, the test statistic is $F = 0.81$ and the *P*-value is 0.374, so there is not sufficient evidence to conclude that the site has an effect on weight. For the column variable of treatment, the test statistic is $F = 7.50$ and the *P*-value is 0.001, so there is sufficient evidence to conclude that the treatment has an effect on weight.

3.   Test statistic: $F = 42.9436$. *P*-value: 0.000. Reject H_0: $\mu_1 = \mu_2 = \mu_3$. There is sufficient evidence to warrant rejection of the claim that the three different types of cigarettes have the same mean amount of tar. Given that the king-size cigarettes have the largest mean of 21.1 mg per cigarette, compared to the other means of 12.9 mg per cigarette and 13.2 mg per cigarette, it appears that the filters do make a difference (although this conclusion is not justified by the results from analysis of variance).

EXCEL

ANOVA

Source of Variation	SS	df	MS	F	P-value	F crit
Between Groups	1083.707	2	541.8533	42.94364	5.29E-13	3.123907449
Within Groups	908.48	72	12.61778			
Total	1992.187	74				

4.   For interaction, the test statistic is $F = 0.8733$ and the P-value is 0.3685, so there does not appear to be an effect from an interaction between gender and whether the subject smokes. For gender, the test statistic is $F = 0.0178$ and the P-value is 0.8960, so gender does not appear to have an effect on body temperature. For smoking, the test statistic is $F = 3.0119$ and the P-value is 0.1082, so there does not appear to be an effect from smoking on body temperature.

EXCEL

ANOVA

Source of Variation	SS	df	MS	F	P-value	F crit
Sample(Gender)	0.005625	1	0.005625	0.017822	0.896012	4.747225
Columns(Smokes)	0.950625	1	0.950625	3.011881	0.108238	4.747225
Interaction	0.275625	1	0.275625	0.873267	0.368476	4.747225
Within	3.7875	12	0.315625			
Total	5.019375	15				

Cumulative Review Exercises

1.   a.   15.5 years, 13.1 years, 22.7 years

   b.   9.7 years, 9.0 years, 18.6 years

   c.   94.5 years2, 80.3 years^2, 346.1 years^2

   d.   Ratio.

2.   Test statistic: $t = -1.383$. Critical values $t = \pm 2.160$ (assuming a 0.05 significance level). (Tech: P-value = 0.1860.) Fail to reject H_0: $\mu_1 = \mu_2$. There is not sufficient evidence to support the claim that there is a difference between the means for the two groups.

MINITAB
Difference = mu (Presidents) - mu (Monarchs)
Estimate for difference: -7.21
95% CI for difference: (-18.33, 3.90)
T-Test of difference = 0 (vs not =): T-Value = -1.38 P-Value = 0.187 DF = 15

3.   Normal, because the histogram is approximately bell-shaped (or the points in a normal quantile plot are reasonably close to a straight-line pattern with no other pattern that is not a straight-line pattern).

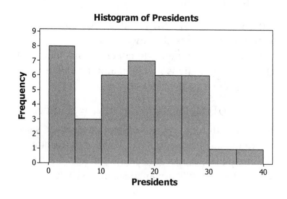

4.   12.3 years $< \mu < 18.7$ years

MINITAB
One-Sample T: Presidents
Variable   N   Mean   StDev   SE Mean   95% CI
Presidents 38   15.50   9.72   1.58   (12.30, 18.70)

5. a. $H_0: \mu_1 = \mu_2 = \mu_3$

    b. Because the P-value of 0.051 is greater than the significance level of 0.05, fail to reject the null hypothesis of equal means. There is not sufficient evidence to warrant rejection of the claim that the three means are equal. The three populations do not appear to have means that are significantly different.

6. a. $r = 0.918$. Critical values: $r = \pm 0.707$. P-value = 0.001. There is sufficient evidence to support the claim that there is a linear correlation between September weights and the subsequent April weights.

    MINITAB
    Correlations: September, April
    Pearson correlation of September and April = 0.918
    P-Value = 0.001

    b. $\hat{y} = 9.28 + 0.823x$

    MINITAB
    Regression Analysis: April versus September
    The regression equation is
    April = 9.3 + 0.823 September
    Predictor   Coef   SE Coef   T   P
    Constant   9.28   11.04   0.84   0.433
    September   0.8227   0.1450   5.67   0.001

    c. $\hat{y} = 9.28 + 0.823(94) = 86.6$ kg, which is not very close to the actual April weight of 105 kg.

7. a. $z = \dfrac{345 - 280}{65} = 1$; $P(z > 1) = 0.1587$.

    b. $z = \dfrac{215 - 280}{65} = -1$; $P(-1 < z < 1) = 0.8413 - 0.1587 = 0.6826$ (Tech: 0.6827)

    c. $z = \dfrac{319 - 280}{65 / \sqrt{25}} = 3$; $P(z < 3) = 0.9987$.

    d. 80th percentile: $x = \mu + z \cdot \sigma = 280 + .84 \cdot 65 = 334.6$ (Tech: 334.7)

8. a. $0.20 \cdot 1000 = 200$ 200

    b. $0.175 < p < 0.225$

    MINITAB
    Test and CI for One Proportion
    Sample   X   N   Sample p   95% CI
    1   200 1000   0.200000   (0.175621, 0.226159)

    c. Yes. The confidence interval shows us that we have 95% confidence that the true population proportion is contained within the limits of 0.175 and 0.225, and 1/4 is not included within that range.

9. a. The distribution should be uniform, with a flat shape. The given histogram agrees (approximately) with the uniform distribution that we expect.

    b. No. A normal distribution is approximately bell-shaped, but the given histogram is far from being bell-shaped.

10. Test statistic: $\chi^2 = 10.400$. Critical value: $\chi^2 = 16.919$ (assuming a 0.05 significance level). P-value > 0.10 (Tech: 0.319). There is not sufficient evidence to warrant rejection of the claim that the digits are selected from a population in which the digits are all equally likely. There does not appear to be a problem with the lottery.

    MINITAB
    Chi-Square Goodness-of-Fit Test for Observed Counts in Variable: C1
    N   DF Chi-Sq P-Value
    200   9   10.4   0.319

Chapter 13: Nonparametric Statistics

Section 13-2

1. The only requirement for the matched pairs is that they constitute a simple random sample. There is no requirement of a normal distribution or any other specific distribution. The sign test is "distribution free" in the sense that it does not require a normal distribution or any other specific distribution.

3. H_0: There is no difference between the populations of September weights and April weights. H_1: There is a difference between the populations of September weights and April weights. The sample data do not contradict H_1 because the numbers of positive signs (2) and negative signs (7) are not exactly the same.

5. The test statistic of $x = 1$ is less than or equal to the critical value of 2 (from Table A-7.) There is sufficient evidence to warrant rejection of the claim of no difference. There does appear to be a difference.

7. The test statistic of $z = \dfrac{(151+0.5) - \frac{1007}{2}}{\sqrt{1007}/2} = -22.18$ falls in the critical region bounded by $z = -1.96$ and

   1.96. There is sufficient evidence to warrant rejection of the claim of no difference. There does appear to be a difference.

9. There are 9 positive signs, 1 negative sign, 0 ties, and $n = 10$. The test statistic of $x = 1$ is less than or equal to the critical value of 1 (from Table A-7). There is sufficient evidence to warrant rejection of the claim of no difference. There does appear to be a difference.

11. The test statistic of $z = \dfrac{(14+0.5) - \frac{32}{2}}{\sqrt{32}/2} = -0.53$ does not fall in the critical region bounded by $z = \pm 1.96$.

    There is not sufficient evidence to warrant rejection of the claim of no difference. There does not appear to be a difference.

    ```
    MINITAB
    Sign Test for Median: Ht - HtOpp-
    Sign test of median = 0.00000 versus not = 0.00000
                N    Below   Equal   Above       P   Median
        C7     34      14      2      18    0.5966   1.000
    ```

13. The test statistic of $z = \dfrac{(52+0.5) - \frac{291}{2}}{\sqrt{291}/2} = -10.90$ is in the critical region bounded by $z = \pm 2.575$. There

    is sufficient evidence to warrant rejection of the claim of no difference. The YSORT method appears to have an effect on the gender of the child. (Because so many more boys were born than would be expected with no effect, it appears that the YSORT method is effective in increasing the likelihood that a baby will be a boy.)

15. The test statistic of $z = \dfrac{(123+0.5) - \frac{280}{2}}{\sqrt{280}/2} = -1.97$ is not in the critical region bounded by $z = \pm 2.575$.

    There is not sufficient evidence to warrant rejection of the claim that the touch therapists make their selections with a method equivalent to random guesses. The touch therapists do not appear to be effective in selecting the correct hand.

17. The test statistic of $z = \dfrac{(12+0.5) - \frac{40}{2}}{\sqrt{40}/2} = -2.37$ is not in the critical region bounded by $z = \pm 2.575$. There

    is not sufficient evidence to warrant rejection of the claim that the median is equal to 5.670 g. The quarters appear to be minted according to specifications.

    ```
    MINITAB
    Sign Test for Median: Post-1964 Quarters
    Sign test of median = 5.670 versus not = 5.670
                           N    Below   Equal   Above       P   Median
    Post-1964 Quarters    40      28      0      12    0.0166   5.636
    ```

19. The test statistic of $z = \dfrac{(1+0.5) - \frac{34}{2}}{\sqrt{34}/2} = -5.32$ is in the critical region bounded by $z = \pm 1.96$. There is sufficient evidence to warrant rejection of the claim that the median amount of Coke is equal to 12 oz. Consumers are not being cheated because they are generally getting more than 12 oz of Coke, not less.

>MINITAB
>Sign Test for Median: CKREGVOL
>Sign test of median = 12.00 versus not = 12.00

	N	Below	Equal	Above	P	Median
CKREGVOL	36	1	2	33	0.0000	12.20

21. Second approach: The test statistic of $z = \dfrac{(30+0.5) - \frac{105}{2}}{\sqrt{105}/2} = -4.29$ is in the critical region bounded by $z = -1.645$, so the conclusions are the same as in Example 4.

Third approach: The test statistic of $z = \dfrac{(38+0.5) - \frac{106}{2}}{\sqrt{106}/2} = -2.82$ is in the critical region bounded by $z = -1.645$, so the conclusions are the same as in Example 4. The different approaches can lead to very different results; as seen in the test statistics of –4.21, –4.29, and –2.82. The conclusions are the same in this case, but they could be different in other cases.

Section 13-3

1. The only requirements are that the matched pairs be a simple random sample and the population of differences be approximately symmetric. There is no requirement of a normal distribution or any other specific distribution. The Wilcoxon signed-ranks test is "distribution free" in the sense that it does not require a normal distribution or any other specific distribution.

3. The sign test uses only the signs of the differences, but the Wilcoxon signed-ranks test uses ranks that are affected by the magnitudes of the differences.

5. Test statistic: $T = 6$. Critical value: $T = 8$. Reject the null hypothesis that the population of differences has a median of 0. There is sufficient evidence to warrant rejection of the claim of no difference. There does appear to be a difference.

>MINITAB
>Test of median = 0.000000 versus median not = 0.000000

	N	N for Test	Wilcoxon Statistic	P	Estimated Median
AGE DIFF	10	10	6.0	0.032	-11.50

7. Convert $T = 247$ to the test statistic $z = \dfrac{247 - \dfrac{32(32+1)}{4}}{\sqrt{\dfrac{32(32+1)(2 \cdot 32+1)}{24}}} = -0.32$.

Critical values: $z = \pm 1.96$. (Tech: P-value = 0.751.) Fail to reject the null hypothesis that the population of differences has a median of 0. There is not sufficient evidence to warrant rejection of the claim of no difference. There does not appear to be a difference.

>MINITAB
>Test of median = 0.000000 versus median not = 0.000000

	N	N for Test	Wilcoxon Statistic	P	Estimated Median
HtDiff	34	32	247.0	0.758	-0.5000

9. Convert $T = 196$ to the test statistic $z = \dfrac{196 - \dfrac{40(40+1)}{4}}{\sqrt{\dfrac{40(40+1)(2 \cdot 40+1)}{24}}} = -2.88$.

Critical values: $z = \pm 2.57$ 5. (Tech: P-value = 0.004.) There is sufficient evidence to warrant rejection of the claim that the median is equal to 5.670 g. The quarters do not appear to be minted according to specifications.

MINITAB
Test of median = 5.670 versus median not = 5.670

	N	N for Test	Wilcoxon Statistic	P	Estimated Median
Post-1964 Quarters	40	40	196.0	0.004	5.638

11. Convert $T = 15.5$ to the test statistic $z = \dfrac{15.5 - \dfrac{34(34+1)}{4}}{\sqrt{\dfrac{34(34+1)(2 \cdot 34+1)}{24}}} = -4.82$.

Critical values: $z = \pm 1.96$. (Tech: P-value = 0.000.) There is sufficient evidence to warrant rejection of the claim that the median amount of Coke is equal to 12 oz. Consumers are not being cheated because they are generally getting more than 12 oz of Coke, not less.

MINITAB
Test of median = 12.000000 versus median not = 12.000000

	N	N for Test	Wilcoxon Statistic	P	Estimated Median
Volume	36	34	15.5	0.000	-0.2000

13. a. Min: 0 and Max: $1 + 2 + \ldots + 74 + 75 = 2850$

b. $\dfrac{2850}{2} = 1425$

c. $2850 - 850 = 2000$

d. $\dfrac{n(n+1)}{2} - k$

Section 13-4

1. Yes. The two samples are independent because the flight data are not matched. The samples are simple random samples. Each sample has more than 10 values.

3. H_0: Arrival delay times from Flights 19 and 21 have the same median. There are three different possible alternative hypotheses: H_1: Arrival delay times from Flights 19 and 21 have different medians. H_1: Arrival delay times from Flight 19 have a median greater than the median of arrival delay times from Flight 21. H_1: Arrival delay times from Flight 19 have a median less than the median of arrival delay times from Flight 21.

5. $R_1 = 137.5$, $R_2 = 162.5$, $\mu_R = \dfrac{12(12+12+1)}{2} = 150$, $\sigma_R = \sqrt{\dfrac{12 \cdot 12(12+12+1)}{12}} = 17.321$, test statistic:

$z = \dfrac{137.5 - 150}{17.321} = -0.72$. Critical values: $z = \pm 1.96$. (Tech: P-value = 0.4705.) Fail to reject the null hypothesis that the populations have the same median. There is not sufficient evidence to warrant rejection of the claim that Flights 19 and 21 have the same median arrival delay time.

7. $R_1 = 253.5$, $R_2 = 124.5$, $\mu_R = \dfrac{13(13+14+1)}{2} = 182$, $\sigma_R = \sqrt{\dfrac{13 \cdot 14(13+14+1)}{12}} = 20.607$, test statistic:

$z = \dfrac{253.5 - 182}{20.607} = 3.47$. Critical values: $z = \pm 1.96$. (Tech: P-value = 0.0005.) Reject the null hypothesis that the populations have the same median. There is sufficient evidence to reject the claim that for those treated with 20 mg of atorvastatin and those treated with 80 mg of atorvastatin, changes in LDL cholesterol have the same median. It appears that the dosage amount does have an effect on the change in LDL cholesterol.

9. $R_1 = 501$, $R_2 = 445$, $\mu_R = \dfrac{22(22+21+1)}{2} = 484$, $\sigma_R = \sqrt{\dfrac{22 \cdot 21(22+21+1)}{12}} = 41.158$, test statistic:

$z = \dfrac{501 - 484}{41.158} = 0.41$. Critical value: $z = 1.645$. (Tech: P-value = 0.3398.) Fail to reject the null hypothesis that the populations have the same median. There is not sufficient evidence to support the claim that subjects with medium lead levels have full IQ scores with a higher median than the median full IQ score for subjects with high lead levels. It does not appear that lead level affects full IQ scores.

11. $R_1 = 2420$, $R_2 = 820$, $\mu_R = \dfrac{40(40+40+1)}{2} = 1620$, $\sigma_R = \sqrt{\dfrac{40 \cdot 40(40+40+1)}{12}} = 103.923$, test

statistic: $z = \dfrac{2420 - 1620}{103.923} = 7.70$. Critical values: $z = \pm 1.96$. (Tech: P-value = 0.0000.) Reject the null hypothesis that the populations have the same median. It appears that the design of quarters changed in 1964.

13.   Using $U = 12 \cdot 11 + \dfrac{12(12+1)}{2} - 123.5 = 86.5$, we get $z = \dfrac{86.5 - \dfrac{12 \cdot 11}{2}}{\sqrt{\dfrac{12 \cdot 11 \cdot (12+11+1)}{12}}} = 1.26$. The test statistic is

the same value with opposite sign.

Section 13-5

1. $R_1 = 1 + 10 + 12.5 + 5 + 8 = 36.5$, $R_2 = 3 + 14 + 15 + 2 + 12.5 + 6 = 52.5$, $R_3 = 7 + 16 + 10 + 4 + 10 = 47$

Low Lead Level	Medium Lead Level	High Lead Level
70 (1)	72 (3)	82 (7)
85 (10)	90 (14)	93 (16)
86 (12.5)	92 (15)	85 (10)
76 (5)	71 (2)	75 (4)
84 (8)	86 (12.5)	85 (10)
	79 (6)	

3. $n_1 = 5$, $n_2 = 6$, $n_3 = 5$, and $N = 5 + 6 + 5 = 16$.

5.   Test statistic: $H = \dfrac{12}{15(15+1)} \left(\dfrac{33^2}{5} + \dfrac{22^2}{5} + \dfrac{65^2}{5} \right) - 3(15+1) = 9.9800$. Critical value: $\chi^2 = 5.991$. (Tech:

P-value = 0.0068.) Reject the null hypothesis of equal medians. The data suggest that the different miles present different levels of difficulty.

7. Test statistic: $H = \dfrac{12}{21(21+1)}\left(\dfrac{86^2}{7}+\dfrac{97^2}{7}+\dfrac{48^2}{7}\right)-3(21+1)=4.9054$. Critical value: $\chi^2=5.991$. (Tech:

   P-value $= 0.0861$.) Fail to reject the null hypothesis of equal medians. The data do not suggest that larger cars are safer.

9. Test statistic: $H = \dfrac{12}{121(121+1)}\left(\dfrac{5277.5^2}{78}+\dfrac{1112^2}{22}+\dfrac{991.5^2}{21}\right)-3(121+1)=8.0115$. Critical value:

   $\chi^2=9.210$. (Tech: P-value $= 0.0182$.) Fail to reject the null hypothesis of equal medians. The data do not suggest that lead exposure has an adverse effect.

11. Test statistic: $H = \dfrac{12}{75(75+1)}\left(\dfrac{1413.5^2}{25}+\dfrac{650.5^2}{25}+\dfrac{786^2}{25}\right)-3(75+1)=27.9098$. Critical value:

   $\chi^2=5.991$. (Tech: P-value: 0.0000.) Reject the null hypothesis of equal medians. There is sufficient evidence to warrant rejection of the claim that the three different types of cigarettes have the same median amount of nicotine. It appears that the filters do make a difference.

13. Using $\Sigma T = 16{,}836$ (see table below) and $N = 25+25+25 = 75$, the corrected value of H is

   $\dfrac{27.9098}{1-\dfrac{16{,}836}{75^3-75}}$ =29.0701, which is not substantially different from the value found in Exercise 11.

   In this case, the large numbers of ties do not appear to have a dramatic effect on the test statistic H.

Nicotine Level	Rank	t	t^3-t
0.2	1.5	2	6
0.6	4.5	2	6
0.7	6.5	2	6
0.8	17.0	19	6840
0.9	28.0	3	24
1.0	33.5	8	504
1.1	48.0	21	9240
1.2	61.0	5	120
1.3	65.5	4	60
1.4	69.0	3	24
1.7	73.5	2	6
		SUM	16,836

Section 13-6

1. The methods of Section 10-3 should not be used for predictions. The regression equation is based on a linear correlation between the two variables, but the methods of this section do not require a linear relationship. The methods of this section could suggest that there is a correlation with paired data associated by some nonlinear relationship, so the regression equation would not be a suitable model for making predictions.

3. r represents the linear correlation coefficient computed from sample paired data; ρ represents the parameter of the linear correlation coefficient computed from a population of paired data; r_s denotes the rank correlation coefficient computed from sample paired data; ρ_s represents the rank correlation coefficient computed from a population of paired data. The subscript s is used so that the rank correlation coefficient can be distinguished from the linear correlation coefficient r. The subscript does not represent the standard deviation s. It is used in recognition of Charles Spearman, who introduced the rank correlation method.

5. $r_s = 1$. Critical values are $r_s = \pm 0.886$ (From Table A-9.) Reject the null hypothesis of $\rho_s = 0$. There is sufficient evidence to support a claim of a correlation between distance and time.

7. $r_s = 0.821$. Critical values: $r_s = \pm 0.786$ (From Table A-9.) Reject the null hypothesis of $\rho_s = 0$. There is sufficient evidence to support the claim of a correlation between the quality scores and prices. These results do suggest that you get better quality by spending more.

    MINITAB
    Pearson correlation of Rank Price and Rank Quality = 0.821

9. $r_s = -0.929$. Critical values: $r_s = \pm 0.786$ (From Table A-9.) Reject the null hypothesis of $\rho_s = 0$. There is sufficient evidence to support the claim of a correlation between the two judges. Examination of the results shows that the first and third judges appear to have opposite rankings.

    MINITAB
    Pearson correlation of First and Second = -0.929

11. $r_s = 1$. Critical values: $r_s = \pm 0.886$ (From Table A-9.) Reject the null hypothesis of $\rho_s = 0$. There is sufficient evidence to conclude that there is a correlation between overhead widths of seals from photographs and the weights of the seals.

    MINITAB
    Pearson correlation of Rank Width and Rank Weight = 1.000

13. $r_s = 0.394$. Critical values: $r_s = \pm \dfrac{1.96}{\sqrt{40-1}} = \pm 0.314$. Reject the null hypothesis of $\rho_s = 0$. There is sufficient evidence to conclude that there is a correlation between the systolic and diastolic blood pressure levels in males.

    MINITAB
    Pearson correlation of RANK SYS and RANK DIAS = 0.394

15. $r_s = 0.651$. Critical values: $r_s = \pm \dfrac{1.96}{\sqrt{48-1}} = \pm 0.286$. Reject the null hypothesis of $\rho_s = 0$. There is sufficient evidence to conclude that there is a correlation between departure delay times and arrival delay times.

    MINITAB
    Pearson correlation of RANK DEPART and RANK ARRIVE = 0.651

17. a. $r_s = \pm \sqrt{\dfrac{2.447^2}{2.447^2 + 8 - 2}} = \pm 0.707$ is not very close to the values of $r_s = \pm 0.738$ found in Table A-9.

    b. $r_s = \pm \sqrt{\dfrac{2.763^2}{2.763^2 + 30 - 2}} = \pm 0.463$ is quite close to the values of $r_s = \pm 0.467$ found in Table A-9.

Section 13-7

1. No. The runs test can be used to determine whether the sequence of World Series wins by American League teams and National League teams is not random, but the runs test does not show whether the proportion of wins by the American League is significantly greater than 0.5.

3. a. Answers vary, but here is a sequence that leads to rejection of randomness because the number of runs is 2, which is very low: W W W W W W W W W W W W E E E E E E E E

   b. Answers vary, but here is a sequence that leads to rejection of randomness because the number of runs is 17, which is very high: W E W E W E W E W E W E W E W E W W W W

5. $n_1 = 19$, $n_2 = 15$, $G = 16$, critical values: 11, 24 (From Table A-10.) Fail to reject randomness. There is not sufficient evidence to support the claim that we elect Democrats and Republicans in a sequence that is not random. Randomness seems plausible here.

7. $n_1 = 20$, $n_2 = 10$, $G = 16$, critical values: 9, 20 (From Table A-10.) Fail to reject randomness. There is not sufficient evidence to reject the claim that the dates before and after July 1 are randomly selected.

9. $n_1 = 24$, $n_2 = 21$, $G = 17$, $\mu_G = \dfrac{2 \cdot 24 \cdot 21}{24 + 21} + 1 = 23.4$, $\sigma_G = \sqrt{\dfrac{2 \cdot 24 \cdot 21(2 \cdot 24 \cdot 21 - 24 - 21)}{(24 + 21)^2 (24 + 21 - 1)}} = 3.3007$.

Test statistic: $z = \dfrac{17 - 23.4}{3.3007} = -1.94$. Critical values: $z = \pm 1.96$. (Tech: P-value = 0.05252.) Fail to reject randomness. There is not sufficient evidence to reject randomness. The runs test does not test for disproportionately more occurrences of one of the two categories, so the runs test does not suggest that either conference is superior.

11. The median is 2453, $n_1 = 23$, $n_2 = 23$, $G = 4$, $\mu_G = \dfrac{2 \cdot 23 \cdot 23}{23 + 23} + 1 = 24$,

$\sigma_G = \sqrt{\dfrac{2 \cdot 23 \cdot 23(2 \cdot 23 \cdot 23 - 23 - 23)}{(23 + 23)^2 (23 + 23 - 1)}} = 3.3553$. Test statistic: $z = \dfrac{4 - 24}{3.3553} = -5.96$. Critical values:

$z = \pm 1.96$. (Tech: P-value = 0.0000.) Reject randomness. The sequence does not appear to be random when considering values above and below the median. There appears to be an upward trend, so the stock market appears to be a profitable investment for the long term, but it has been more volatile in recent years.

13. a. No solution provided.

   b. The 84 sequences yield these results: 2 sequences have 2 runs, 7 sequences have 3 runs, 20 sequences have 4 runs, 25 sequences have 5 runs, 20 sequences have 6 runs, and 10 sequences have 7 runs.

   c. With P(2 runs) = 2/84, P(3 runs) = 7/84, P(4 runs) = 20/84, P(5 runs) = 25/84, P(6 runs) = 20/84, and P(7 runs) = 10/84, each of the G values of 3, 4, 5, 6, 7 can easily occur by chance, whereas $G = 2$ is unlikely because P(2 runs) is less than 0.025. The lower critical value of G is therefore 2, and there is no upper critical value that can be equaled or exceeded.

   d. Critical value of $G = 2$ agrees with Table A-10. The table lists 8 as the upper critical value, but it is impossible to get 8 runs using the given elements.

Chapter Quick Quiz

1. Distribution-free test

2. 57 has rank $\dfrac{1 + 2 + 3}{3} = 2$, 58 has rank 4, and 61 has rank 5.

3. The efficiency rating of 0.91 indicates that with all other factors being the same, rank correlation requires 100 pairs of sample observations to achieve the same results as 91 pairs of observations with the parametric test for linear correlation, assuming that the stricter requirements for using linear correlation are met.

4. The Wilcoxon rank-sum test does not require that the samples be from populations having a normal distribution or any other specific distribution.

5. $G = 4$

6. Because there are only two runs, all of the values below the mean occur at the beginning and all of the values above the mean occur at the end, or vice versa. This indicates an upward (or downward) trend.

7. Sign test and Wilcoxon signed-ranks test

8. Rank correlation

9. Kruskal-Wallis test

10. Test claims involving matched pairs of data; test claims involving nominal data; test claims about the median of a single population

Review Exercises

1. The test statistic of $z = \dfrac{(44+0.5)-\frac{106}{2}}{\sqrt{106}/2} = -1.65$ is not less than or equal to the critical value of $z = -1.96$. Fail to reject the null hypothesis of $p = 0.5$. There is not sufficient evidence to warrant rejection of the claim that in each World Series, the American League team has a 0.5 probability of winning.

2. There are 6 positive signs, 0 negative signs, 0 ties, and $n = 7$. The test statistic of $x = 0$ is less than or equal to the critical value of 0. There is sufficient evidence to reject the claim of no difference. It appears that there is a difference in cost between flights scheduled 1 day in advance and those scheduled 30 days in advance. Because all of the flights scheduled 30 days in advance cost less than those scheduled 1 day in advance, it is wise to schedule flights 30 days in advance.

3. The test statistic of $T = 0$ is less than or equal to the critical value of 0. There is sufficient evidence to reject the claim that differences between fares for flights scheduled 1 day in advance and those scheduled 30 days in advance have a median equal to 0. Because all of the flights scheduled 1 day in advance have higher fares than those scheduled 30 days in advance, it appears that it is generally less expensive to schedule flights 30 days in advance instead of 1 day in advance.

   ```
   MINITAB
   Test of median = 0.000000 versus median not = 0.000000
                               N for   Wilcoxon            Estimated
                        N      Test    Statistic    P      Median
   One Day – 30 Days  7        7         28.0     0.022     357.8
   ```

4. The sample mean is 54.8 years. $n_1 = 19$, $n_2 = 19$, and the number of runs is $G = 18$. The critical values are 13 and 27 (From Table A-10.) Fail to reject the null hypothesis of randomness. There is not sufficient evidence to warrant rejection of the claim that the sequence of ages is random relative to values above and below the mean. The results do not suggest that there is an upward trend or a downward trend.

5. $r_s = 0.714$. Critical values: $r_s = \pm0.738$ (From Table A-9.) Fail to reject the null hypothesis of $\rho_s = 0$. There is not sufficient evidence to support the claim that there is a correlation between the student ranks and the magazine ranks. When ranking colleges, students and the magazine do not appear to agree.

   ```
   MINITAB
   Pearson correlation of Student Ranks and USNEWS Ranks = 0.714
   ```

6. The test statistic of $z = \dfrac{(13+0.5)-\frac{32}{2}}{\sqrt{32}/2} = -0.88$ is not in the critical region bounded by $z = \pm1.96$. There is not sufficient evidence to warrant rejection of the claim that the population of differences has a median of zero. Based on the sample data, it appears that the predictions are reasonably accurate, because there does not appear to be a difference between the actual high temperatures and the predicted high temperatures.

7. Convert $T = 230.5$ to the test statistic $z = \dfrac{230.5 - \dfrac{32(32+1)}{4}}{\sqrt{\dfrac{32(32+1)(2\cdot 32+1)}{24}}} = -0.62$.

   Critical values: $z = \pm1.96$. (Tech: P-value = 0.531.) There is not sufficient evidence to warrant rejection of the claim that the population of differences has a median of zero. Based on the sample data, it appears that the predictions are reasonably accurate, because there does not appear to be a difference between the actual high temperatures and the predicted high temperatures.

7. (continued)

    MINITAB
    Test of median = 0.000000 versus median not = 0.000000

		N for	Wilcoxon		Estimated
	N	Test	Statistic	P	Median
TEMP	35	32	230.5	0.537	-0.5000

8. Test statistic: $H = \dfrac{12}{27(27+1)}\left(\dfrac{91^2}{9} + \dfrac{112.5^2}{9} + \dfrac{174.5^2}{9}\right) - 3(27+1) = 6.6305$.

    Critical value: $\chi^2 = 5.991$. (Tech: P-value = 0.0363.) Reject the null hypothesis of equal medians. Interbreeding of cultures is suggested by the data.

9. $R_1 = 60$, $R_2 = 111$, $\mu_R = \dfrac{9(9+9+1)}{2} = 85.5$, $\sigma_R = \sqrt{\dfrac{9 \cdot 9(9+9+1)}{12}} = 11.3248$, test statistic:

    $z = \dfrac{60-85.5}{11.3248} = -2.25$. Critical values: $z = \pm 1.96$. (Tech: P-value = 0.0243.) Reject the null hypothesis that the populations have the same median. Skull breadths from 4000 b.c. appear to have a different median than those from a.d. 150.

10. $r_s = 0.473$. Critical values: $r_s = \pm 0.587$. Fail to reject the null hypothesis of $\rho_s = 0$. There is not sufficient evidence to support the claim that there is a correlation between weights of plastic and weights of food.

    Pearson correlation of Rank Plastic and Rank Rood = 0.473

Cumulative Review Exercises

1. $\bar{x} = 14.6$ hours, median = 15.0 hours, $s = 1.7$ hours, $s^2 = 2.9$ hour^2, range = 6.0 hours

2. a. Convenience sample

    b. Because the sample is from one class of statistics students, it is not likely to be representative of the population of all fulltime college students.

    c. Discrete

    d. Ratio

3. The data meet the requirement of being from a normal distribution. H_0: $\mu = 14$ hours . H_1: $\mu > 14$ hours .

    Test statistic: $t = \dfrac{14.55-14}{1.701/\sqrt{20}} = 1.446$. Critical value: $t = 1.729$ (assuming a 0.05 significance level). P-value > 0.05 (Tech: 0.0822). Fail to reject H_0. There is not sufficient evidence to support the claim that the mean is greater than 14 hours.

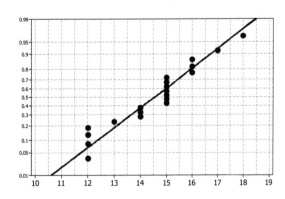

4. The test statistic of $x = 5$ is not less than or equal to the critical value of 4 (from Table A-7.) There is not sufficient evidence to support the claim that the sample is from a population with a median greater than 14 hours.

5. 13.8 hours $< \mu < 15.3$ hours. We have 95% confidence that the limits of 13.8 hours and 15.3 hours contain the true value of the population mean.

$$14.55 - 2.101 \cdot \frac{1.701}{\sqrt{20}} < \mu < 14.55 + 2.101 \cdot \frac{1.701}{\sqrt{20}}$$

6. $r = 0.205$. Critical values: $r = \pm 0.811$. P-value = 0.697. There is not sufficient evidence to support the claim of a linear correlation between price and quality score. It appears that you don't get better quality by paying more.

    MINITAB
    Pearson correlation of Price and Quality = 0.205
    P-Value = 0.697

7. $r_s = -0.543$. Critical values: $r_s = \pm 0.886$. Fail to reject the null hypothesis of $\rho_s = 0$. There is not sufficient evidence to support the claim that there is a correlation between price and rank.

    MINITAB
    Pearson correlation of Rank and Price Rank = -0.543
    P-Value = 0.266

8. $0.276 < p < 0.343$. Because the value of 0.25 is not included in the range of values in the confidence interval, the result suggests that the percentage of all such telephones that are not functioning is different from 25%.

$$\frac{229}{740} - 1.96\sqrt{\frac{\left(\frac{229}{740}\right)\left(\frac{511}{740}\right)}{740}} < p < \frac{229}{740} + 1.96\sqrt{\frac{\left(\frac{229}{740}\right)\left(\frac{511}{740}\right)}{740}}$$

9. $n = \dfrac{\left[z_{\alpha/2}\right]^2 (0.25)}{E^2} = \dfrac{1.645^2 (0.25)}{0.02^2} = 1692$ (Tech: 1691)

10. There must be an error, because the rates of 13.7% and 10.6% are not possible with samples of size 100.

Chapter 14: Statistical Process Control

Section 14-2

1. No. If we know that the manufacture of quarters is within statistical control, we know that the three out-of-control criteria are not violated, but we know nothing about whether the specification of 5.670 g is being met. It is possible to be within statistical control by manufacturing quarters with weights that are very far from the desired target of 5.670 g.

3. To use an $\bar{x}$ chart without an R chart is to ignore variation, and amounts of variation that are too large will result in too many defective goods or services, even though the mean might appear to be acceptable. To use an R chart without an $\bar{x}$ chart is to ignore the central tendency, so the goods or services might not vary much, but the process could be drifting so that daily process data do not vary much, but the daily means are steadily increasing or decreasing.

5. $\bar{\bar{x}} = 267.11$ lb, $\bar{R} = 54.96$ lb, $n = 7$.

   For the R chart: $\text{LCL} = D_3\bar{R} = 0.076 \cdot 54.96 = 4.18$ lb and $\text{UCL} = D_4\bar{R} = 1.924 \cdot 54.96 = 105.74$ lb.

   For the $\bar{x}$ chart: $\text{LCL} = \bar{\bar{x}} - A_2\bar{R} = 267.11 - 0.419 \cdot 54.96 = 244.08$ lb and

   $\quad\text{UCL} = \bar{\bar{x}} + A_2\bar{R} = 267.11 + 0.419 \cdot 54.96 = 290.14$ lb.

7. The R chart does not violate any of the out-of-control criteria, so the variation of the process appears to be within statistical control

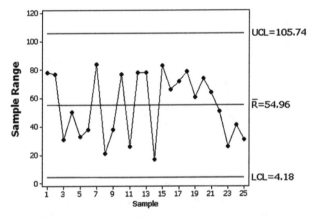

9. $\bar{\bar{x}} = 14.250°C$, $\bar{R} = 0.414°C$, $n = 10$.

   For the R chart: $\text{LCL} = D_3\bar{R} = 0.223 \cdot 0.414 = 0.092°C$ and $\text{UCL} = D_4\bar{R} = 1.777 \cdot 0.414 = 0.736°C$.

   For the $\bar{x}$ chart: $\text{LCL} = \bar{\bar{x}} - A_2\bar{R} = 14.250 - 0.308 \cdot 0.414 = 14.122°C$ and

   $\quad\text{UCL} = \bar{\bar{x}} + A_2\bar{R} = 14.250 + 0.308 \cdot 0.414 = 14.377°C$.

11. Because there is a pattern of an upward trend and there are points lying beyond the control limits, the $\bar{x}$ chart shows that the process is out of statistical control.

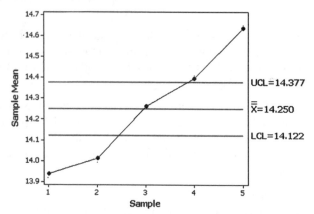

13. $\bar{s} = 0.0823$ g, $n = 5$. The R chart and the s chart are very similar in their pattern.
    $LCL = B_3\bar{s} = 0 \cdot 0.0823 = 0$ g and $UCL = B_4\bar{s} = 2.089 \cdot 0.0823 = 0.1719$ g .

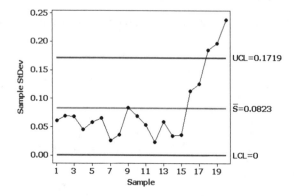

Section 14-3

1. No, the process does not appear to be within statistical control. There is a downward trend, there are at least 8 consecutive points all lying above the centerline, and there are at least 8 consecutive points all lying below the centerline. Because the proportions of defects are decreasing, the manufacturing process is not deteriorating; it is improving.

3. LCL denotes the lower control limit. Because the value of –0.000025 is negative and the actual proportion of defects cannot be less than 0, we should replace that value by 0.

5. The process appears to be within statistical control.

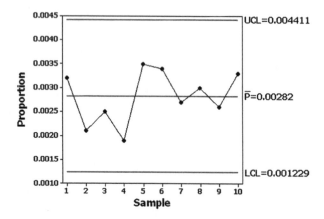

7. $\bar{p} = 0.01407$, LCL $= \bar{p} - 3\sqrt{\dfrac{\overline{pq}}{n}} = 0.01407 - 3\sqrt{\dfrac{(0.01407)(0.98593)}{100,000}} = 0.012953$

$$\text{UCL} = \bar{p} + 3\sqrt{\dfrac{\overline{pq}}{n}} = 0.01407 + 3\sqrt{\dfrac{(0.01407)(0.98593)}{100,000}} = 0.015187$$

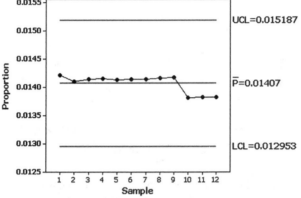

Because there appears to be a pattern of a downward shift and there are at least 8 consecutive points all lying above the centerline, the process is not within statistical control.

9. $\bar{p} = 0.55231$, LCL $= \bar{p} - 3\sqrt{\dfrac{\overline{pq}}{n}} = 0.55231 - 3\sqrt{\dfrac{(0.55231)(0.44769)}{1000}} = 0.50513$

$$\text{UCL} = \bar{p} + 3\sqrt{\dfrac{\overline{pq}}{n}} = 0.55231 + 3\sqrt{\dfrac{(0.55231)(0.44769)}{1000}} = 0.0.59948$$

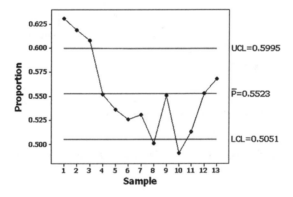

The process is out of control because there are points lying beyond the control limits and there are at least 8 points all lying below the centerline. The percentage of voters started to increase in recent years, and it should be much higher than any of the rates shown.

The process appears to be within statistical control. Ideally, there would be an upward trend due to increasing rates of college enrollments among high school graduates.

11. $\bar{p} = 0.0268$, LCL $= \bar{p} - 3\sqrt{\dfrac{\overline{pq}}{n}} = 0.0268 - 3\sqrt{\dfrac{(0.0268)(0.9732)}{500}} = 0.00513$

UCL $= \bar{p} + 3\sqrt{\dfrac{\overline{pq}}{n}} = 0.0268 + 3\sqrt{\dfrac{(0.0268)(0.9732)}{500}} = 0.04847$

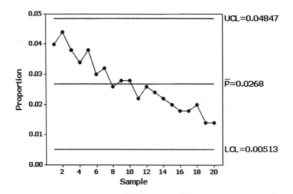

There is a pattern of a downward trend and there are at least 8 consecutive points all below the centerline, so the process does not appear to be within statistical control. Because the rate of defects is decreasing, the process is actually improving and we should investigate the cause of that improvement so that it can be continued.

13. $n\bar{p} = 10{,}000 \cdot 0.00126 = 12.6$, LCL $= n\bar{p} - 3\sqrt{n\overline{pq}} = 12.6 - 3\sqrt{10{,}000(0.00126)(0.99874)} = 1.9578$

UCL $= n\bar{p} + 3\sqrt{n\overline{pq}} = 12.6 + 3\sqrt{10{,}000(0.00126)(0.99874)} = 23.2422$

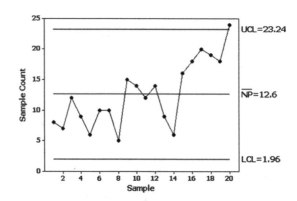

Except for the vertical scale, the control chart is identical to the one obtained for Example 1.

Chapter Quick Quiz

1. Process data are data arranged according to some time sequence. They are measurements of a characteristic of goods or services that result from some combination of equipment, people, materials, methods, and conditions.

2. Random variation is due to chance, but assignable variation results from causes that can be identified, such as defective machinery or untrained employees.

3. There is a pattern, trend, or cycle that is obviously not random. There is a point lying outside of the region between the upper and lower control limits. There are at least 8 consecutive points all above or all below the centerline.

4. An R chart uses ranges to monitor variation, but an $\bar{x}$ chart uses sample means to monitor the center (mean) of a process.

5. No. The R chart has at least 8 consecutive points all lying below the centerline and there are points lying beyond the upper control limit. Also, there is a pattern showing that the ranges have jumped in value for the most recent samples.

6. $\bar{R}$ = 52.8 ft. In general, a value of $\bar{R}$ is found by first finding the range for the values within each individual subgroup; the mean of those ranges is the value of $\bar{R}$.

7. No. The $\bar{x}$ chart has a point lying below the lower control limit.

8. $\bar{\bar{x}}$ = 3.95 ft. In general, a value of $\bar{\bar{x}}$ is found by first finding the mean of the values within each individual subgroup; the mean of those subgroup means is the value of $\bar{\bar{x}}$.

9. A p chart is a control chart of the proportions of some attribute, such as defective items.

10. Because there is a downward trend, the process is not within statistical control, but the rate of defects is decreasing, so we should investigate and identify the cause of that trend so that it can be continued.

Review Exercises

1. $\bar{\bar{x}}$ = 2781.71 kWh, $\bar{R}$ = 1729.38 kWh, n = 6.
    For the R chart: $\text{LCL} = D_3\bar{R} = 0 \cdot 1729.38 = 0$ kWh and
    $\qquad \text{UCL} = D_4\bar{R} = 2.004 \cdot 1729.38 = 3465.678$ kWh .
    For the $\bar{x}$ chart: $\text{LCL} = \bar{\bar{x}} - A_2\bar{R} = 2781.71 - 0.483 \cdot 1729.38 = 1946.419$ kWh and
    $\qquad \text{UCL} = \bar{\bar{x}} - A_2\bar{R} = 2781.71 - 0.483 \cdot 1729.38 = 3617.001$ kWh .

2. $\bar{R}$ = 1729.4 kWh, n = 6. The process variation is within statistical control.
    $\text{LCL} = D_3\bar{R} = 0 \cdot 1729.4 = 0.0000$ kWh and $\text{UCL} = D_4\bar{R} = 2.004 \cdot 1729.4 = 3465.7176$ kWh .

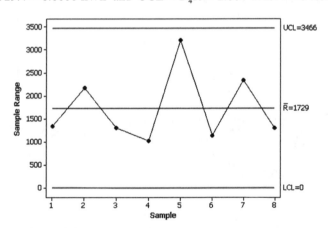

3. $\bar{\bar{x}}$ = 2781.71 kWh, $\bar{R}$ = 1729.4 kWh, n = 6. The process mean is within statistical control.
    $\text{LCL} = \bar{\bar{x}} - A_2\bar{R} = 2781.71 - 0.483 \cdot 1729.4 = 1946.4098$ kWh and
    $\text{UCL} = \bar{\bar{x}} + A_2\bar{R} = 2781.71 + 0.483 \cdot 1729.4 = 3617.0102$ kWh .

3.   (continued)

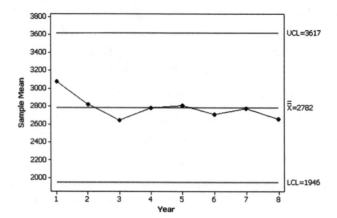

4.   There does not appear to be a pattern suggesting that the process is not within statistical control. There is 1 point that appears to be exceptionally low. (The author's power company made an error in recording and reporting the energy consumption for that time period.)

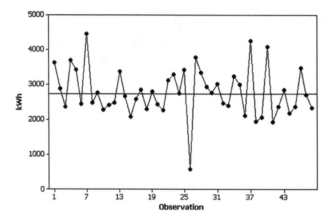

5. $\bar{p} = 0.056$, $\text{LCL} = \bar{p} - 3\sqrt{\dfrac{\bar{p}\bar{q}}{n}} = 0.056 - 3\sqrt{\dfrac{(0.056)(0.944)}{100}} = -0.01298$; use 0

$\text{LCL} = \bar{p} + 3\sqrt{\dfrac{\bar{p}\bar{q}}{n}} = 0.056 + 3\sqrt{\dfrac{(0.056)(0.944)}{100}} = 0.12498$

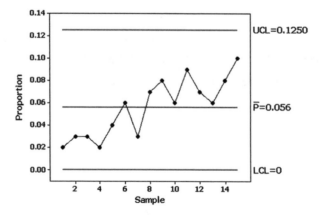

Because there are 8 consecutive points above the centerline and there is an upward trend, the process does not appear to be within statistical control.

Cumulative Review Exercises

1. $0.519 < p < 0.581$. Because all of the values in the confidence interval estimate of the population proportion are greater than 0.5, it does appear that the majority of adults believe that it is not appropriate to wear shorts at work.

   ```
   MINITAB
   Sample   X    N  Sample p      95% CI
   1      550  1000  0.550000  (0.518557, 0.581148)
   ```

2. a. $1 - 0.55 = 0.45$

   b. $(0.55)^5 = 0.0503$

   c. $1 - (0.55)^5 = 0.950$

3. $r = 0.820$. Critical values: $r = \pm 0.602$. P-value = 0.00202. There is sufficient evidence to support the claim that there is a linear correlation between yields from regular seed and kiln-dried seed. The purpose of the experiment was to determine whether there is a difference in yield from regular seed and kiln-dried seed (or whether kiln-dried seed produces a higher yield), but results from a test of correlation do not provide us with the information we need to address that issue.

   ```
   MINITAB
   Pearson correlation of Regular and Kiln-dried = 0.820
   P-Value = 0.002
   ```

4. H_0: $\mu_d = 0$. H_1: $\mu_d < 0$. Test statistic: $t = -1.532$. Critical value: $t = -1.812$ (assuming a 0.05 significance level). P-value < 0.05 (Tech: 0.0783). Fail to reject H_0. There is not sufficient evidence to support the claim that kiln-dried seed is better in the sense that it produces a higher mean yield than regular seed. (The sign test can be used to arrive at the same conclusion; the test statistic is $x = 3$ and the critical value is 1. Also, the Wilcoxon signed-ranks test can be used; the test statistic is $T = 13.5$ and the critical value is 8.)

   ```
   Minitab
   Paired T for Regular - Kiln-dried
   95% upper bound for mean difference: 0.200
   T-Test of mean difference = 0 (vs < 0): T-Value = -1.53
   P-Value = 0.078
   ```

5. For the sample of yields from regular seed, $\bar{x} = 20.0$ and for the sample of yields from kiln-dried seed, $\bar{x} = 21.0$, so there does not appear to be a significant difference. For the sample of yields from regular seed, s = 3.4 and for the sample of yields from kiln-dried seed, s = 4.1, so there does not appear to be a significant difference.

6. $\bar{p} = 0.122$, $\text{LCL} = \bar{p} - 3\sqrt{\dfrac{\bar{p}\bar{q}}{n}} = 0.122 - 3\sqrt{\dfrac{(0.122)(0.878)}{50}} = -0.01686$; use 0

$\text{LCL} = \bar{p} + 3\sqrt{\dfrac{\bar{p}\bar{q}}{n}} = 0.122 + 3\sqrt{\dfrac{(0.122)(0.878)}{50}} = 0.26086$

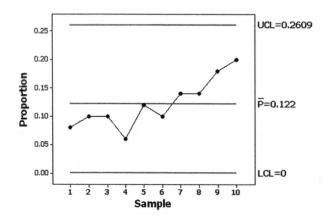

There appears to be a pattern of an upward trend, so the process is not within statistical control.

7.

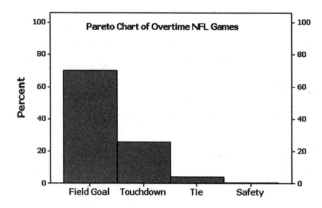

8.   a. $z = \dfrac{17 - 15.2}{2.5} = 0.72$; $P(z > 0.72) = 23.58\%$. With 23.58% of males with head breadths greater than

17 cm, too many males would be excluded.

   b.   5th percentile: $x = \mu + z \cdot \sigma = 15.2 - 1.645 \cdot 2.5 = 11.08$ cm

        95th percentile: $x = \mu + z \cdot \sigma = 15.2 + 1.645 \cdot 2.5 = 19.3$ cm

9.   With a voluntary response sample, the subjects decide themselves whether to be included. With a simple random sample, subjects are selected through some random process in such a way that all samples of the same size have the same chance of being selected. A simple random sample is generally better for use with statistical methods.

10.   Sampling method (part c)